Influencing Customer Demand

Influencing Customer Demand

An Operations Management Approach

Edited by
M. Hemmati and Mohsen S. Sajadieh

CRC Press
Taylor & Francis Group
Boca Raton London New York

CRC Press is an imprint of the
Taylor & Francis Group, an **informa** business

Library of Congress Cataloging-in-Publication Data

Names: Hemmati, M., editor. | Sajadieh, Mohsen S., editor.
Title: Influencing customer demand: an operations management approach /
    edited by M. Hemmati and Mohsen S. Sajadieh.
Description: First edition. | Boca Raton: CRC Press, 2021. | Includes
    bibliographical references and index.
Identifiers: LCCN 2021001510 (print) | LCCN 2021001511 (ebook) | ISBN
    9780367619985 (hardback) | ISBN 9781003107446 (ebook)
Subjects: LCSH: Supply and demand. | Marketing—Management. | Sales
    management. | Production management.
Classification: LCC HB820.I54 2021 (print) | LCC HB820 (ebook) | DDC
    658.8/343—dc23
LC record available at https://lccn.loc.gov/2021001510
LC ebook record available at https://lccn.loc.gov/2021001511

ISBN: 978-0-367-61998-5 (hbk)
ISBN: 978-0-367-62005-9 (pbk)
ISBN: 978-1-003-10744-6 (ebk)

Typeset in Times
by Apex CoVantage, LLC

Contents

Preface

In today's competitive markets, balancing demand and supply is crucial to maximizing revenue and improving customer satisfaction. Demand management as a supply chain management (SCM) process and sales-and-operations planning step can increase efficiency and productivity. It can also increase a firm's ability to satisfy customer demand and balance it with supply to help managers have a successful business.

Influencing demand as one of demand management's modules is a repetitive process that employs techniques such as marketing, selling, promotion, and pricing strategies to increase/decrease demand, shape it, and change product/service mix. It can also alter the timing of demand and its volume. Influencing demand tries to affect customers and convince them to purchase in such a way that supports the company's goals. It also equips the company to predict what and when customers buy a product/service beforehand.

In addition to methods such as marketing, which has received much attention in the literature, many other exogenous and endogenous factors can influence demand. Identifying these factors and the related mathematical models is critical to obtaining the best product/service mix, delivery time, batch size, and so on. This book gives a comprehensive view of influencing factors on customer demand from an operations management perspective, proposes the different forms of demand functions to obtain optimal policies, and discusses their pros and cons. Moreover, applications of these factors and their functions in different industries and real problems are addressed.

This book can help companies measure demand dependency on a variety of factors to make more revenue and gain notable improvements compared to their competitors. It also leads managers to better align supply and demand by affecting the demand shaping and understanding the crucial role of demand management within their SCM. Hence, managers can reduce the related operational costs, including but not limited to inventory holding, deterioration, and shipment costs, while increasing profit, customer satisfaction, and service level.

All in all, this book can be useful for researchers, practitioners, and professionals in academic institutions and industries who have concerns about managing demand and need a comprehensive resource of different aspects of factors that affect demand from conceptual framework to optimization methodology. Moreover, graduate students and teachers may also find this book valuable to augment their current knowledge.

Finally, the editors appreciate all the contributors' time, effort, and dedication toward the successful completion of this book. The editors are also grateful for the editorial assistance of CRC Press/Taylor & Francis, especially the support and cooperation of Erin Harris, Senior Editorial Assistant, and Cindy Renee Carelli, Executive Editor. We enjoyed the experience of working on this book and hope that readers find the book interesting and valuable.

M. Hemmati
Mohsen S. Sajadieh
Tehran, Iran

Editors

M. Hemmati is a PhD student at Amirkabir University of Technology. She received her BS degree in Industrial Engineering from Isfahan University of Technology, Isfahan, Iran, in 2014 and her MSc degree in Industrial Engineering from Amirkabir University of Technology, Tehran, Iran, in 2016. Her research interests include mathematical optimization, supply chain management, pricing, and demand management. Moreover, she has published some papers in scientific conferences and journals and received Iran's national elite scholarship award. In 2017, She received the best MSc theses award from Amirkabir University of Technology.

Mohsen S. Sajadieh is an assistant professor in the Industrial Engineering & Management Systems Department at Amirkabir University of Technology, Iran. He earned his PhD in Industrial Engineering from Sharif University of Technology, Iran, in 2009. He held a visiting position at Aarhus School of Business in Denmark in 2008. His research interests center on logistics and supply chain management. Mohsen is an expert, working as a consultant and project manager of logistics and supply chain projects for well-known companies in Iran. He also serves as a referee for many international journals.

1 Introduction

Carlos Castro-Zuluaga and
Mariana Arboleda-Florez
University Eafit

CONTENTS

Demand is the engine of any company and its supply chain. Despite having access to more information and technological tools than ever before, many companies still use statistical forecasts as their sole method for planning their finances, sales, and operations. By doing so, they do not consider other factors that affect and influence the future demand, which usually translates into unfeasible plans.

Conversely, demand management is a means to implement best practices as well as evolved forecasting methods. Demand management includes different factors that influence the demand while reducing the uncertainty of demand forecasts. Consequently, all the other plans of the company are positively impacted. As a result, the company can assemble a demand management model that will comprise an integrated business management model.

Influencing demand is part of the demand management process, which in turn is a fundamental step in the Sales and Operations Planning (S&OP) process. APICS defines S&OP as: "the process that provides management the ability to strategically direct its business to achieve competitive advantage on a continuous basis by integrating customer-focused marketing plans for new and existing products with the management of the supply chain" (Palmatier and Crum 2002).

Although S&OP was created in the early 1980s as one of the main pillars of supply chain management, companies have recently begun to recognize S&OP as one of the main managerial tools that facilitates the executives gaining a wider vision of the near future of the company from a marketing, finance, and operations perspective (Tuomikangas and Kaipia 2014).

By enacting S&OP, companies seek to level both demand and capacity across several stages throughout their supply process. The result of this process is a consensus-based plan, which establishes how the company will meet the customer's requirements. By doing so, the company acquires the capacity to anticipate customer demand, thus achieving greater efficiency in inventory levels and operational costs (Bower 2006; T. Wallace 2010).

1.1 THE S&OP PROCESS

S&OP is a multi-functional and comprehensive monthly process that compiles all the business plans that would traditionally be scattered across the functional areas of product development, operations, logistics, finance, and marketing and sales into a single, integrated model of operation. This hierarchy is presented in Figure 1.1.

S&OP requires the commitment of the company's managerial levels, under the coordination of the CEO, to generate aligned plans that balance the supply and in

FIGURE 1.1 Multi-functional S&OP process.

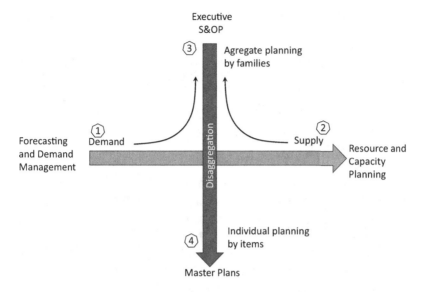

FIGURE 1.2 Components of the S&OP process.

the middle term. This eases the decision-making process when choosing the best combination of markets, products, and customers to be served.

S&OP is carried out and updated in a monthly cycle at an aggregate level of product families. Then, after a disaggregation process, it becomes the detailed plan of procurement, supply, outsourcing, production, warehousing, distribution, and logistics for the near-to-intermediate term.

In the first steps, S&OP treats the potential demand by establishing different product families at the aggregate level for further forecasting of the amount of sales for each family. These quantities are expressed in a plan that defines the different resources and possible strategies needed to meet the demand requirements (Wallace and Robert 2008). Figure 1.2 shows the main components of S&OP and their flows of information, which go from 1 to 4.

S&OP planners must be aware of the importance of balancing the middle-term demand and supply at an aggregate level. This balancing should be aligned with the upcoming individual planning, as this is the core of S&OP.

1.2 STEPS OF SALES & OPERATIONS PLANNING

Several authors have defined a number of steps to describe S&OP; furthermore, the steps are quite similar among themselves (Palmatier and Crum 2002; T. F. Wallace and Stahl 2006; Sheldon 2006). Figure 1.3 presents the five steps in which S&OP can be summarized along with the departments, areas, and personnel that are typically involved in each of the steps.

Steps 1–3 are crucial to preparing for the successful completion of the final two steps. As presented in Figure 1.3, it is necessary to have cross-functional collaboration in every step, which ultimately allows for having all the main areas of the

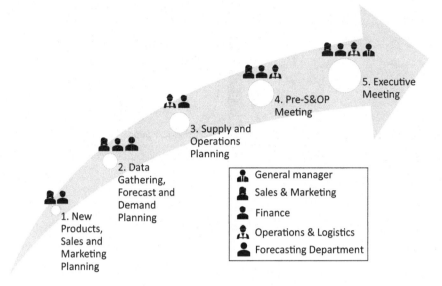

FIGURE 1.3   Steps of the S&OP and involved areas.

company working together at the last step. A brief description of each step is presented below.

1.2.1 Step 1: New Products, Marketing, and Sales Planning

At this step, the appointed team reviews the product portfolio and discusses innovations, new products, and marketing and sales strategies. They also define all the actions related to launching new products, offering promotions, planning special events, giving volume discounts, rethinking price strategies, attracting new customers, and changing competitive strategy, among others. These actions are also updated in a timeline of the demand plan. Then, all those plans will become the primary input for the second step of S&OP. Furthermore, additional information collected in this step, such as competitor evaluations, marketing research, customers panels and feedbacks, etc., must also be shared with the demand planner to ensure the success of the following step.

1.2.2 Step 2: Data Gathering, Forecasting, and Demand Planning

Demand planning is the process that aims to estimate the expected future requirements with the highest possible level of accuracy. Whether it uses the information coming from the orders placed by customers or extrapolated sales history, the demand plan should be adjusted according to the available information from the demand-influencing factors. As mentioned for the previous step, planning should be carefully studied and comprehended before enacting it.

The first activities in the demand planning step consist of gathering and cleaning data for quickly obtaining updated baseline statistical forecasts for the product families or services offered by the company. After these statistical forecasts have been updated, the next step consists of adjusting them through a "collaborative forecasting process," where different demand-influencing factors are analyzed and further included for obtaining a conciliation and recommended demand plan. This will be the main input for the following steps of S&OP. It is highly recommended to have a decision support system when carrying out these activities.

Nowadays, demand planning has evolved to become a more comprehensive process called demand management. This process not only includes planning but also communicating, influencing, managing, and prioritizing the demand to have a more accurate and real demand forecast (Crum and Palmatier 2003).

Within the S&OP literature, demand planning is mostly considered to be a responsibility of the marketing and sales department on a "black box" process, which should be later communicated to the other areas and agreed upon when it needs modifications. Interestingly enough, after surveying different companies, the Institute of Business Forecasting and Planning found that the demand forecasting and planning process resides in different functional areas of the company. Out of surveyed companies, 35% place this process in Operations/Logistics, 27% in Sales and Marketing, 14% in the Forecasting Department, 10% in other areas of the company, 7% in Strategic Planning, and 7% in Finance. This survey shows that there is not a clear consensus on the area in charge of the process across the evaluated companies. This book acknowledges the influence of those who are not directly involved with sales and marketing but who also take part in building up the brand with the customers, as a result of whole operational decisions and business planning.

1.2.3 STEP 3: SUPPLY AND OPERATION PLANNING

In the Supply and Operation Planning step, all the plans related to sourcing, procurement, manufacturing, inventory, warehousing, inbound and outbound transportation, and third-party logistics services are defined and updated for the short and middle term according to the collaborative demand plan established in the previous step. The objective of this step is to obtain both definitive and alternative feasible plans to meet the demand plan, which can be subject to different internal and external constraints. In this step, the plans are developed, assessed, and aggregated by the product family through rough-cut resource planning to obtain the lowest possible cost.

1.2.4 STEP 4: PRE-S&OP MEETING

In this step, called the pre-meeting, middle managers, from all the main areas involved and affected by the S&OP process, attend a meeting in which all the demand, supply, and operations plans are presented for obtaining a definitive consensus (if possible). In this meeting, the attendants from all areas of the company have the opportunity to influence the decisions. The result of this meeting is the agenda for the executive meeting, where all the final decisions that reached a consensus are presented. For those cases where the consensus was not possible, it is necessary to develop scenarios and propose recommendations for the manager to make an informed decision. In any case, the financial and budget effects of the agreed-upon decisions as well as the proposed alternative solutions are evaluated.

1.2.5 STEP 5: EXECUTIVE MEETING

This is the final step of the monthly S&OP process, where the definitive sales and operations plans are approved by the CEO or director and the senior managers of the company who are involved in the process. In this executive meeting, the members review the major decisions made in the previous step and the key performance indicators (KPI) of the company. With this assessment, they try to remove constraints and make decisions about all the points that were not agreed upon in the previous step. They also make a final evaluation of the proposed suggestions and scenarios to verify how the definitive plans impact the budget and strategy of the company (Palmatier and Crum 2002).

Consequently, one of the main outputs of the S&OP process is the consensus-based sales plan of the company. Here, through the demand management process, statistical forecasts are merged with the activities and factors that can impact and influence the demand to obtain a more realistic plan, which aims to achieve the objectives set by the company.

1.3 INFLUENCING THE DEMAND PROCESS

Even though there are some exogenous factors that influence customer demand, most of them are self-inflicted and respond to actual decisions made inside the company. The remainder of this book will present how all the actions that occur in the supply chain can positively or negatively influence the demand. Even more, when companies understand that all processes are both clients and suppliers for the other processes within the company, they realize how important it is to fulfill internal brand promises as well as build internal trust. Only then, when different processes trust one another, will the company be able to work toward the greater goal and build both trust and engagement for their external customer.

Having this in mind, influencing demand should be understood as a process in which each decision should be defined, measured, analyzed, improved, and controlled regardless of the decision introducing a quality change, rearranging the production schedule, changing reorder frequency, or advertising a special offer. Figure 1.4 illustrates the DMAIC process applied to influencing the demand, which is key to ensuring the best possible and reproducible results.

1.4 INFLUENCING CUSTOMER DEMAND FACTORS

Influencing the demand has been a task entrusted, almost exclusively, to the departments of sales and marketing, which should try, by means of different strategies, to convince and capture new and current customers for acquiring preexisting and new products and services in agreement with the goals and objectives defined by the company.

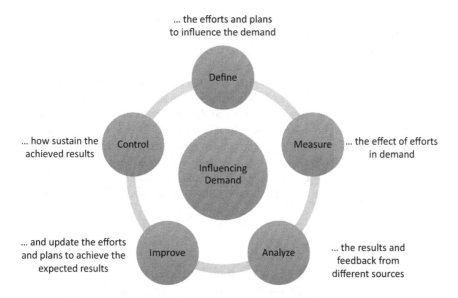

... the efforts and plans
to influence the demand

Define

... how sustain the
achieved results — Control

Influencing
Demand

Measure — ... the effect of efforts
in demand

... and update the efforts
and plans to achieve the
expected results — Improve

Analyze — ... the results and
feedback from
different sources

FIGURE 1.4 The DMAIC process applied to the demand influence process.

The remainder of this chapter presents a brief description of the demand-influencing factors addressed in Chapters 2–13, keeping in mind that these factors involve all the areas in the company as they can directly or indirectly influence the demand. It is important to point out that each chapter also discusses the managerial, mathematical, and conceptual framework of the different demand-influencing factors despite not being mentioned in this introduction.

1.4.1 CHAPTER 2. PRICING POLICIES

Considering that the price can be modified at any stage in the supply chain, basing decisions solely on price could cause a negative impact in both up- and downstream processes. In one scenario, a lower price becomes attractive enough for customers to the point that upstream stages will not be prepared to respond to the higher demand; therefore, a stockout could result from this drop in price. Conversely, a higher price may cause a reduction in the demand, generating an overstock.

Nevertheless, the impact that pricing policies by themselves have on demand is controlled both by the product's price elasticity and the buyer's knowledge of future pricing. For example, if a buyer knows in advance when an increase or decrease of the price will take place, they can decide on whether to overstock on products before prices go up or to stop acquiring them and change their supplier.

1.4.2 CHAPTER 3. ON-SHELF INVENTORY

Retail customers are highly influenced by different factors inside the store. There is a general consensus on the way customers are influenced by how visible the products are and how interestingly (or not) they are displayed on the shelves.

Eye-level items usually get more attention than those that require some effort to be found. However, shelf space is limited, and "eye-level" does not mean the same for all customers; hence, on-shelf inventory becomes a planning problem in distributing the products to be displayed among a retail store's scarce shelf space. Specifically, this embraces the following questions: how much shelf space should each product be assigned and at which vertical shelf level? What products should be placed next to one another, and how should they be horizontally distributed (Baron, Berman, and Perry 2011; Bianchi-Aguiar et al. 2020)? It is necessary to consider these questions when influencing customer demand for maximizing store revenue. This approach faces some difficult challenges that have been largely discussed in the literature, for example, trying to avoid losing sales due to out-of-stock products while maintaining product variety along with a sense of innovation and urgency for the clients.

1.4.3 CHAPTER 4. REBATE CONTRACTS

In the manufacturing industry, there is a frequent operation in which the customer receives a partial refund. This partial refund is known as a rebate, and it is commonly based on a portion of the price the customer paid under specific terms defined by the seller. In operation management, there are several types of rebates. This chapter covers some specific forms of rebates, including mail-in rebates, consumer rebates, wholesale price rebates, and channel or retailer rebates.

Mail-in rebate is a pricing mechanism where a delayed incentive is offered, either through cash or a gift card, by the retailer to the final customer for the purchase of a product (Geng and Mallik 2011); furthermore, in this rebate, customers must meet certain requirements to get the refund (e.g., filling out a form, gathering documentation, sending in the request for the rebate). A consumer rebate is an incentive offered as a refund or a coupon from the manufacturer to the final customer for each unit they buy (Aydin and Porteus 2015). The wholesale price rebate is a strategy wherein the manufacturer offers a discount to the retailer for accepting delivery during an earlier period (Huang, Kuo, and Lu 2014). Channel rebates are payments made from a manufacturer to a retailer based on their sales to final costumers (Taylor 2002); moreover, this rebate always requires a contract, which could be linear- or volume-based while including important information concerning other demand-influencing factors, such as marketing and sales efforts.

This chapter also considers the length of credit period, which refers to a period of time that the supplier grants to the company so the latter can pay for their acquired inventory after the agreed-upon period. The length of the credit period will represent an economic benefit for the supplier only if an additional profit can be generated by increasing the sales volume to offset the incurred opportunity cost. On the other hand, the company can use the credit period to obtain revenue, which they can use to cover their expenses while gaining a profit. Ultimately, this strategy improves the company's cash flow. Depending on the circumstances, the credit period represents benefits for both the supplier and the company; it is important to highlight that the length of the credit period along with some other factors should be agreed upon by both parties. For example, if a manufacturer sells a price-sensitive product to a retailer, they need to coordinate not only the length of the credit period but also a suggested selling price to not disturb the market. Therefore, the length of the credit period can be one of the many non-evident factors that influence the demand.

1.4.4 Chapter 5. Service Level Effects

In general, service level both defines the standard of performance that is expected from a service provider and allows the comparison when formulating a performance indicator. Consequently, each company could choose a definition that suits their industry, whether it is measured in product units, in hours spent fixing a machine, or in minutes waiting on hold. Regarding supply chain management, however, the cycle service level is the expected probability of not hitting a stockout during the next replenishment cycle; in other words, it is the probability of not losing sales. For this reason, service level is usually matched to the safety stock, and companies plan their replenishment cycles and inventory based on the service level they want to achieve.

While a 100% service level might appear desirable, it is usually not a feasible option. Nonetheless, having a low service level is unappealing to potential clients and could be the reason to lose even the most loyal customers. To reduce that risk, most companies plan for different strategies that allow them to reduce the impact of not complying with their service level. Most of those strategies include paying penalties or reducing participation on future sales; these strategies are considered negative consequences for not fulfilling customers' expectations. Nevertheless, other companies have decided to make their service level a part of their added value as a main strategy to attract and retain clients; by doing so, they change the customers' perspective on a low service level while expecting to engage with them.

1.4.5 Chapter 6. Marketing Decisions and Efforts

According to Marketing Insiders Group, "demand is the foundation of all marketing and sales processes, and all businesses looking to grow need to implement effective demand generation strategies" (Rivard 2017). These strategies range from communication strategies, promotions, advertising, and, recently, even hiring social media influencers to generate conversations around the product. However, it is important to deeply understand the business, so these strategies can be aligned with the company's internal processes to avoid unwanted results.

For instance, the positive effect of one creative promotion could be overshadowed by the negative effect of having stockouts, hence resulting in the loss of customer loyalty. At the same time, the operational efforts to achieve the best quality product could be "wasted" if marketers do not choose to communicate it to their clients. Demonstrating that when the company lacks integration between the departments of marketing and sales and operations, it is possible that either the marketing and sales department makes an unfeasible offer or the operations department makes decisions to optimize the manufacturing processes without considering the effect on the customer.

1.4.6 CHAPTER 7. A COMPANY'S REPUTATION

Even though the concept of reputation does not have a unanimous definition due to its abstract nature, it is one of the most important intangible assets of a company. The reputation of a company is often indistinguishable from its brand and perception. Usually, people with a good reputation can easily influence their environment; similarly, it has been demonstrated that companies with a good reputation can influence customers to prefer their products, be less price sensitive, and be more willing to find good attributes on their products' features (Burke, Dowling, and Wei 2018). This translates to a higher awareness of either the leather details or the greener areas when looking for a car from a reputable brand or an MBA from an elite university, respectively.

The reputation of a company involves every stage of the supply chain. From buying low-quality materials to having questionable hiring practices, each step of the supply chain is ultimately responsible for affecting the reputation, thus influencing customers. Nowadays, these kinds of policies have become more accessible to the public due to the spread of the Internet, which makes information widely and readily available. Consequently, people are constantly gathering information about companies; nevertheless, when a person forms an opinion, they usually have not accessed all the information available; they frequently do so by reading online reviews or listening to what other people have to say about their product of interest. Having an integrated business plan should take into consideration how the company wants to be perceived by their customers and the public.

1.4.7 CHAPTER 8. CONGESTION IN THE SYSTEM

Any stage of the supply chain is a system where different inputs go through a transformation process to become a desired output. However, resources are always limited, and sometimes their demand is higher than their capacity, leading to a congested system.

Sometimes, the demand of products or services depends on the congestion of the systems to which they belong; therefore, the decision of whether or not a customer joins the system is based on how congested it is. For instance, in queueing systems, comprised of one or multiple servers, the service rate of customers depends on the congestion of the system. Consequently, the demand will be positively influenced if the customer chooses to stay in the system and actually acquire the products or services. Conversely, the demand will be negatively influenced if the customer chooses to leave the system after perceiving that the congestion surpasses their tolerance threshold. To model how the congestion factor affects the demand, it is necessary to evaluate customers' perceptions of the benefits of acquiring the service or product pondered against their willingness to invest resources toward the acquisition. Therefore, the customer only joins the system if the benefits weigh more than the expected cost of obtaining the products or services (e.g., access fees and waiting costs) (Randhawa 2013; Whitt 2003). This chapter deeply discusses factors such as lead time, capacity sizing, mean waiting time, delay probability, server utilization, and response time on demand, among others.

1.4.8 CHAPTER 9. DYNAMIC INNOVATION CAPABILITIES

According to Michael Porter, one of the ways companies must differentiate themselves from other companies is through innovation. This allows them to achieve a competitive advantage by influencing current and future demand. To do so, every stage of the supply chain has to develop innovation capabilities to identify, develop, and implement original solution-oriented actions for addressing new or previously unsolved problems (Teece, Pisano, and Shuen 1997). There are two main factors that drive companies to be innovative; the first one is that current markets are highly dynamic, and the second is that customers are continuously demanding innovative products and services that exceed their expectations and satisfy their needs. Companies that seek differentiation by innovation must heavily invest in research and development. In this way, they guarantee a successful process for developing new products or services that meet customers' expectations. Nevertheless, the time needed to research and develop the service or product, its expected life cycle, the marketing and launching plans, among others are important factors that also play an important role in the success or failure of the innovation attempt.

The ability to align, adapt, and reconfigure resources, processes, products, and systems to the market and customers' needs are a requirement that both the firm and its supply chain must be capable of in a dynamic way for successfully meeting changing markets and customer demand, which, in turn, is an important factor that innovative companies use to influence current and future demand.

1.4.9 CHAPTER 10. DEMAND SENSITIVITY TO TIME

Whether it is for natural or business reasons, there are many products whose demand is considered time-sensitive. Two of the most common perspectives to understanding sensitivity are freshness and seasonality. Freshness refers to the way customers perceive the products in terms of how recently they were collected. Traditionally, freshness has been associated with perishable produce like fruits and vegetables; however, in recent years, electronic products and fast-fashion clothing have also been associated with freshness, given that they heavily rely on a sense of novelty to be purchased. Fresh products are more appealing to customers and could expect constant demand levels based on their historical data, as long as they remain fresh; from that point on, demand is expected to decline. Nonetheless, the product remains saleable until it begins to deteriorate, driving researchers to focus on replenishment and pricing policies (e.g., by lowering price through time) to reduce losing sales (Wu et al. 2016; Banerjee and Agrawal 2017).

On the contrary, seasonal products are associated with products sold during specific times of the year, such as Christmas, Halloween, or back-to-school. However, some companies evidence seasonality during the day, such as restaurants, gyms, or health centers. A common practice is that companies self-induce seasonality when they are prone to having an end-of-quarter rush or special anniversary sales. Moreover, seasonality differs from freshness on how the company prepares in advance for the demand peaks after recognizing the seasonal patterns. This practice helps the company avoid losing sales and potential customers (Banerjee and Sharma 2010).

1.4.10 Chapter 11. Inflation-Dependent Demand

Many economies, markets, and countries face fluctuating inflation rates that directly impact the operation of the companies. In such cases, companies can use their net working capital to buy inventory and prevent potential losses due to price fluctuation. For instance, it is better to acquire more products in advance than to buy a smaller quantity with the same money later.

There are several approaches for modeling how inflation affects demand. These models are mainly used to decide about pricing policies and optimal replenishment quantities, which influence the demand (Tripathi 2011). Additionally, there are approaches in the literature that include other conditions, such as inventory loss, deteriorating, credit period, and cash flow, among others. These conditions are useful when constructing models to predict the effects of inflation on demand, some of which will be deeply discussed in this chapter.

1.4.11 Chapter 12. Substitute and Complementary Goods

The effects of the price of substitutes/complementary goods on demand have been discussed in some academic works (Walters 1991). Substitute goods are defined as products from different brands or manufactures that are perceived as similar among themselves and may be bought in replacement of one another. Therefore, substitute product demands are interrelated, given that customers can indistinctly switch between them based on their perceived advantages when purchasing. Although there are many factors that can impact the demand of substitutes, their inventory level and selling price are two of the main factors that influence their demand. However, this impact is easier to notice when the substitute products have cross-price elasticity (i.e., a price increase of product A results in a demand increase of substitute product B alongside a demand decrease of product A and vice versa) (Andreyeva, Long, and Brownell 2010). There are different types of substitutes, such as perfect and nonperfect, gross and net, and within and cross-category substitutes, which are critical when modeling the impact they have on the demand.

Conversely, complementary goods are products or services that are sold individually but used together; moreover, the demand of one product generates demand for its complementary products. Once more, the sale price and inventory level of the complementary goods are two of the main factors that impact the demand. Furthermore, customers tend to buy complementary goods when they are simultaneously available and at least one of the product's prices is attractive enough; therefore, there is a positive cross-price elasticity effect. However, the opposite effect occurs when the price of complementary goods increases.

1.4.12 Chapter 13. Mass and Social Media

Social media first appeared in 2000. However, it was only 10 years ago when it became the huge industry it is today with hundreds of millions of users worldwide. Social media has reshaped society in a way such that one person could easily access any piece of information they want. However, they are only shown the information that is algorithmically considered most relevant for them. Consequently, companies have found the most powerful tool for targeting advertising, building relationships, and improving customer segmentation as clients actively choose to see their content, connect with the company, and refer other friends by clicking a button.

Moreover, social media highly influences businesses' online reputation, and it is one of the biggest sources of online reviews. While this means reaching more potential buyers than ever before, it also means to be in continuous scrutiny and on the edge of "going viral" for the wrong reasons. One of the biggest retail stores in the United States and the largest car manufacturer in the world have experienced social media backlash due to marketing strategies, and they are just one example of how a 20-second video, a picture, or a blog post could cause a millionaire significant losses.

1.4.13 CHAPTER 14. PAST AND FUTURE OF DEMAND FORECASTING MODELS

Up until now, we have reviewed several demand-influencing models and factors. However, the factors presented in this book do not cover the full extent of the state of the art; moreover, there are many other considerations when managing demand from an S&OP perspective that should be studied. One of them consists of the forecasting methods the company chooses to implement once it has decided on the demand-influencing factors it wants to model. Chapter 14 categorizes quantitative forecasting methods based on demand dependency as independent and dependent demand models.

For independent demand, different time-series demand models are presented. Those models are classified according to the time-series pattern. For stationary time series that do not present seasonality nor any evident trend, models such as naïve, simple average (SA), Simple Moving Average (SMAn), Weighted Moving Average (WMAn,w), and Single Exponential Smoothing (SESα) are reviewed. Then, for time series without seasonality and with a trend, different variations of the Double Exponential Smoothing (DESα,β) model are reviewed. Furthermore, for time series with seasonality, decomposition models, Triple Exponential Smoothing is presented.

Conversely, for dependent demand, the authors review techniques for estimating the parameters of the causal models, which are modeled in terms of some influential factors, including willingness-to-pay dependent, product-greenness dependent, product-visibility dependent, refund dependent, and service quality–dependent demand models. These models can be either univariate and multivariate regression models or simulation-based models. More advanced techniques, referring to artificial intelligence and machine learning are also discussed.

Additionally, another important consideration consists of the decision over the measure of the forecasting performance, which plays an important role in the forecasting process due to its convenience for comparing errors from different models and calibrating them. This chapter discusses some of the most common forecast performance measures used by both practitioners and academics, which are mean absolute error (MAE), mean absolute percentage error (MAPE), mean square error (MSE), and mean absolute percentage error (MAPE) (Armstrong and Collopy 1992; Hyndman and Koehler 2006). Finally, the authors also review demand information sharing and, specifically, advance demand information (ADI) in both forms of perfect and imperfect information; to conclude with several directions for future research.

To summarize, this introductory chapter has introduced the factors that affect and influence the demand, it has also introduced how the demand-planning step links these factors to the S&OP. Additionally, this chapter emphasizes how the company's success lies in balancing the supply and demand in the middle term while integrating both the internal demand management process with the supply and financial planning processes.

This chapter also explains how influencing demand is a process that marketing, sales, and other areas of the company perform to convince customers to acquire products or services in a way that supports the objectives of the company and how there are many traditional factors that affect customer demand, such as promotions, discounts, price strategies, and advertisements, among others. However, there are other factors, such as rebate contracts, congestion in the system, the length of the credit period, inflation, etc., that directly or indirectly impact the demand. Hence, it is important not only for the demand planner but also for the supply and financial planners to know how all these factors influence the demand and how they impact the different plans of the company.

Each chapter of this book addresses one of the influencing factors mentioned in this introduction, showing its conceptual framework, demand functions, advantages and disadvantages in its implementation, examples, and case studies, allowing the reader to have a better understanding of how other factors can influence demand, in addition to the traditional ones.

We strongly believe this book can help middle and top managers in the main functional areas of a company to know how many factors influence demand and to understand the underlying role of demand management in their company's planning. This knowledge can be applied throughout the supply chain. Therefore, managers can work to reduce all the logistic costs including ordering, carrying, stockout, and shipping costs while they seek to increase the profit throughout an improvement of availability and service levels. By doing so, they can improve customer satisfaction. Finally, the goal is to maintain and improve the competitiveness of the company and the supply chain, achieving the objectives set by the CEOs and shareholders.

REFERENCES

Andreyeva, Tatiana, Michael W. Long, and Kelly D. Brownell. 2010. The impact of food prices on consumption: A systematic review of research on the price elasticity of demand for food. *American Journal of Public Health* 100, no. 2: 216–222.

Armstrong, J. Scott, and Fred Collopy. 1992. Error measures for generalizing about forecasting methods: Empirical comparisons. *International Journal of Forecasting* 8, no. 1: 69–80.

Aydin, Goker, and Evan L. Porteus. 2015. Manufacturer-to-retailer versus manufacturer-to-consumer rebates in a supply chain. In *Retail Supply Chain Management*, 349–386. Boston: Springer.

Banerjee, Snigdha, and Swati Agrawal. 2017. Inventory model for deteriorating items with freshness and price dependent demand: Optimal discounting and ordering policies. *Applied Mathematical Modelling* 52: 53–64.

Banerjee, Snigdha, and Ashish Sharma. 2010. Optimal procurement and pricing policies for inventory models with price and time dependent seasonal demand. *Mathematical and Computer Modelling* 51, no. 5–6: 700–714.

Baron, Opher, Oded Berman, and David Perry. 2011. Shelf space management when demand depends on the inventory level. *Production and Operations Management* 20, no. 5: 714–726.

Bianchi-Aguiar, Teresa, Alexander Hübner, Maria Antónia Carravilla, and José F. Oliveira. 2020. Retail shelf space planning problems: A comprehensive review and classification framework. *European Journal of Operational Research* 289, no. 1: 1–16.

Bower, Patrick. 2006. How the S&OP process creates value in the supply chain. *Journal of Business Forecasting* 25, no. 2: 20.

Burke, Paul F., Grahame Dowling, and Edward Wei. 2018. The relative impact of corporate reputation on consumer choice: Beyond a halo effect. *Journal of Marketing Management* 34, no. 13–14: 1227–1257.

Crum, Colleen, and George E. Palmatier. 2003. *Demand Management Best Practices: Process, Principles, and Collaboration.* Boca Raton: J. Ross Publishing.

Geng, Qin, and Suman Mallik. 2011. Joint mail-in rebate decisions in supply chains under demand uncertainty. *Production and Operations Management* 20, no. 4: 587–602.

Huang, Kwei-Long, Chia-Wei Kuo, and Ming-Lun Lu. 2014. Wholesale price rebate vs. Capacity expansion: The optimal strategy for seasonal products in a supply chain. *European Journal of Operational Research* 234, no. 1: 77–85.

Hyndman, Rob J., and Anne B. Koehler. 2006. Another look at measures of forecast accuracy. *International Journal of Forecasting* 22, no. 4: 679–688.

Palmatier, George E., and Colleen Crum. 2002. *Enterprise Sales and Operations Planning: Synchronizing Demand, Supply and Resources for Peak Performance.* Boca Raton: J. Ross Publishing.

Randhawa, R. S. 2013. Accuracy of fluid approximations for queueing systems with congestion-sensitive demand and implications for capacity sizing. *Operations Research Letters* 41, no. 1: 27–31.

Rivard, Johanna. 2017. What demand marketing really means—marketing insider group. May 23. https://marketinginsidergroup.com/content-marketing/demand-marketing-really-means/.

Sheldon, Donald H. 2006. *World Class Sales & Operations Planning: A Guide to Successful Implementation and Robust Execution.* Fort Lauderdale, FL: J. Ross Publishing.

Taylor, Terry A. 2002. Supply chain coordination under channel rebates with sales effort effects. *Management Science* 48, no. 8: 992–1007.

Teece, David J., Gary Pisano, and Amy Shuen. 1997. Dynamic capabilities and strategic management. *Strategic Management Journal* 18, no. 7: 509–533.

Tripathi, R. P. 2011. Optimal pricing and ordering policy for inflation dependent demand rate under permissible delay in payments. *International Journal of Business, Management and Social Sciences* 2, no. 4: 35–43.

Tuomikangas, Nina, and Riikka Kaipia. 2014. A coordination framework for sales and operations planning (S&OP): Synthesis from the literature. *International Journal of Production Economics* 154: 243–262.

Wallace, Thomas F., and A. Robert Stahl. 2006. *Sales & Operations Planning: The Executive's Guide; Balancing Demand and Supply; Aligning Units and Enhancing Teamwork.* Cincinnati, OH: TF Wallace & Company.

Wallace, Thomas F., and A. Robert Stahl. 2008. *Sales & Operations Planning-the How-to Handbook.* Cincinnati, OH: TF Wallace & Company.

Wallace, Tom. 2010. Executive sales & operations planning: Cost and benefit analysis. *Journal of Business Forecasting* 29, no. 3.

Walters, Rockney G. 1991. Assessing the impact of retail price promotions on product substitution, complementary purchase, and interstore sales displacement. *Journal of Marketing* 55, no. 2: 17–28.

Whitt, Ward. 2003. How multiserver queues scale with growing congestion-dependent demand. *Operations Research* 51, no. 4.

Wu, Jiang, Chun-Tao Chang, Mei-Chuan Cheng, Jinn-Tsair Teng, and Faisal B. Al-Khateeb. 2016. Inventorymanagement for fresh produce when the time-varying demand depends on product freshness, stock level and expiration date. *International Journal of Systems Science: Operations & Logistics* 3, no. 3: 138–147.

2 Pricing Policies

Zeinab Vosooghi and M. Hemmati
Department of Industrial Engineering, Amirkabir
University of Technology, Tehran, Iran

CONTENTS

2.1 INTRODUCTION

Pricing, which can be assumed as one of the most critical factors in retailers' profit, can strongly affect the behavior of consumers (Duan and Ventura 2020). It is also the only component of marketing that makes revenue (LaPlaca 1997). The price flexibility and price changes in today's marketing world are increasingly noticeable. In this regard, a reduction in the span of standard prices and a short changing-price period for online stores are presented by Cavallo (2018) and Gorodnichenko and Talavera (2017), respectively. So, retailers have to face more serious challenges to set the prices in this competitive market, and a rational pricing strategy is needed to ease making pricing decisions and achieving a noted benefit (Hinterhuber and Bertini 2011). Determining a profitable pricing strategy for firms is dependent on such different factors as the retailers' objectives, the manner of consumers, and the pricing situation (Sajadieh and Jokar 2009 and Hemmati, Fatemi Ghomi, and Sajadieh 2017).

As mentioned in Lilien, Kotler, and Moorthy (1992), there are five steps for customers to complete their purchase cycle: experiencing arousal by internal and external motivations, searching for satisfying brands, making a comparison between different products, making the decision regarding the purchase, and encountering post-purchase feelings. Consequently, since pricing strategies directly affect all five steps, pricing decisions can dramatically influence firms' profits by changing the demand level. The inability of decision-makers in achieving enough information regarding the demand function makes them rely on nonoptimal pricing strategies since making these decisions is so difficult in this circumstance. Thus, simultaneous cooperation of operations and marketing areas, which may influence practical decisions of the supply chain, can help the companies achieve a higher level of profitability.

In recent years, numerous studies investigated the effects of different marketing behaviors on customer demand. Huang (2013) stated that since the role of an item's price on affecting a customer's evaluation of that product is so critical, pricing is the most important factor in changing retailer demand. This effect, called customer price elasticity, can be either positive—like fashion products—or negative—like non-fashion ones (Lichtenstein, Ridgway, and Netemeyer 1993). Generally, the more price elasticity a customer has, the cheaper product she/he will prefer to buy. It also can be changed by taking different pricing techniques, which affects both the firm's demand and others' profit directly (Seyedhosseini et al. 2019). So, price-dependent demand models are the most frequently applied functions in previous studies.

The remainder of this chapter is organized as follows: initially, we present the conceptual framework of the proposed factors, then a review of the demand function models will be presented. Different practical applications of the proposed models, some related examples, and instances of the optimization models are then investigated. Research trends will be reviewed, and finally, a conclusion and research suggestions for future studies will be presented.

2.2 CONCEPTUAL FRAMEWORK

One of the most critical decisions in supply chain management is pricing (Monroe 2002), so, explaining its different characteristics seems quite essential and is undertaken hereafter.

There are three managerial levels for pricing strategies: industry, market, and transactional strategies (Marn and Rosiello 1992). The top and most common level of price management is industry pricing, which is directly changed by any changes in demand, supply, or costs. The main goal here is finding the current and upcoming "market tone" to establish the best pricing strategy (Baker, Marn, and Zawada 2010). Market strategy's objective is specifying all customers' determining factors for buying a product, their reference price, the way they compare prices with others, and the relative benefit they make from their purchase. By doing so, retailers can set the best price for their products by considering their competitors' strategies. The last level of price management is the way sellers can manage the price of each transaction. This includes determining the base price, discounts, auctions, and incentives (Marn and Rosiello 1992).

Considering pricing strategies, three main categories of cost-based, customer value-based, and competitor-oriented pricing can be mentioned as the major ones (Jampala 2016), which are explained in detail here.

1. Cost-based pricing: As Kotler and Armstrong (2010) mentioned, in this strategy, retailers use the production, distribution, selling, and risk-related return rates to determine prices. It includes such different strategies as cost-plus (markup) pricing, breakeven (target-return) pricing, absorption-cost pricing, and marginal-cost pricing. Adding a fixed percentage to the unit cost and ignoring the competitors' pricing strategy as well as customer demand determine the markup pricing strategy. In breakeven pricing, product prices are set by target-return cost, which is calculated by finding the zero-profit point (dividing the item's fixed cost to its variable cost). The third strategy—absorption-cost pricing—determines the selling prices by adding fixed cost, variable cost, administration cost, and the required arbitrary margin with each other. Setting the price by considering all direct variable costs and some parts of fixed costs can also be mentioned as the marginal-cost pricing strategy (Jampala 2016). There are two important limitations for this strategy: ignoring the market condition as well as competitors' status in decision-making and imposing more financial pressure on customers due to the selling-price increase caused by firms' lost motivation for reducing production costs.
2. Value-based pricing: In this strategy, the added value of the product to customers, rather than the production cost or historical prices, is assumed as the main factor for price setting (Cressman 2012). According to Ingenbleek et al. (2003), providing an exact evaluation of products' value for customers is the main defect of this method, which is the most common policy in the related literature as well.

    The diagrammatic depiction of cost-based and value-based pricing strategies are shown in Figure 2.1 (Kotler and Armstrong 2010).
3. Competitor-oriented pricing: This strategy uses parity, premium, and discount pricing as the main policies. For using these options, the competitor's price is considered as the basic price level (Blythe 2006), and setting the prices equal to, more, or less than the benchmark determines the premium, discount, and parity policy, respectively.

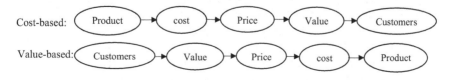

FIGURE 2.1   Schematic representation of value- and cost-based pricing.

There are some other pricing strategies, some of which are presented here:

4. Dynamic pricing, which is also called time-based, or surge pricing, is a pricing policy in which real-time demand and supply (which varies over time) determine a product's price (Kotler and Armstrong 2010). Airlines, hotel rooms, and ride-sharing prices are some evident examples of applying this policy to real-world problems. To categorize dynamic pricing models, Elmaghraby and Keskinocak (2003) proposed price-posted and price-discovery models in which prices are set by the company and customers, respectively, as two main categories.

5. Differential pricing: In this strategy, which is one of the most important parts of dynamic pricing, sellers determine the item's price for each customer independently; that is, deciding on one product's selling price is based on different user segments (Brassington and Pettitt 2006). In some special circumstances, it can be basically similar to the discriminatory pricing as well.

6. Product-line pricing: This policy, which is also known as the product-line promotion method of pricing, sets the selling price of a production line's products in such a way that the price of the line—as a whole—will be optimized (Jampala 2016). To do so, companies use both the cost differences between various products of the line and their prices in competitor companies to determine different price steps between them.

7. Psychological pricing: Considering psychological factors, rather than economic ones, in setting prices makes the fundamental concept of this strategy (Kotler and Armstrong 2010). Since it is obvious that just-below prices affect customers' emotions and persuade them to buy the products, by selecting this strategy, firms increase their sales by influencing customers' perceptions and increasing their willingness to buy (Faith and Edwin 2014). Using $0.99 rather than $1 is one of the most common applications of this policy in selling products.

8. Odd-even pricing is another psychological tactic that uses odd prices (ending in odd numbers) to influence customers' perceptions. By using under-even prices, sellers can psychologically create a sense of discount in customers, which can directly affect their desire to buy products and consequently influence the firm's profit (Faith and Edwin 2014).

9. Tender pricing: Firms use competitive bidding from customers to set the products' prices in this policy. This strategy is usually used in business organizations.

10. By-product pricing: In this policy, selling by-products can help the company recover some of the essential costs and reduce the main product's price. Reducing the price of the main petroleum products, achieved from a crude oil refining process, by selling their by-product, ethylene, as one of the main components of all plastic goods is one obvious instance of this strategy.

11. Bundle pricing: Setting a single price for a multi-product package (Farese, Kimbrell, and Woloszyk 2001). Obviously, to encourage customers to buy more, the bundle price has to be less than the aggregated price of all included-in-a-package products.

12. Target pricing: In this strategy, investment rate of return is assumed as the main criteria for price setting. This policy, which is mostly used in competitive industries, can be implemented through three main steps: (1) determining the selling price of a product in such a way that guarantees its survival in the competitive market, (2) defining the desirable profit margin, and (3) calculating the maximum allowable production cost by subtracting the profit margin from the selling price (Jampala 2016).

13. Prestige pricing, which is also named image pricing, is setting a product's price at a high level for conveying a sense of brilliant quality to target customers. As mentioned by Cannon and Morgan (1990), cheap prices in some products prevent customers from buying them since it indicates low quality, so to increase their profit, sellers have to determine their prices are high enough to stop the customers from worrying about the quality of their products and encourage them to buy the high-quality products (Brassington and Pettitt 2006).

14. Affordability-based pricing: This strategy, which is occasionally accompanied by governmental subsidy policies (known as social welfare pricing), is used for setting the price of essential commodities, which make up the majority of market demand. This strategy's objective is setting the commodities' price independent of their production cost in such a way that the basic needs of most demand markets are fully satisfied (Jampala 2016).

15. Predatory pricing (undercutting pricing), one of consumers' favorite policies, is setting extremely low prices for items such that other firms cannot compete with them. This policy is often used by firms to eliminate their competitors to monopolize the market (Brassington and Pettitt 2006). Bankrupting competitors is another goal of newcomers, which encourages them to determine set prices lower than the production cost (Blythe 2006).

New product pricing strategies are also listed as follows:

1. Online pricing: The Internet has drastically influenced pricing in obtaining information and performing transactions (Ratchford 2009). Regarding the differences between online and offline markets, Baye et al. (2007) stated that ease of information collection by both competitors and online consumers can be mentioned as the most noticeable one. This explosive volume of information makes markets competitive by providing customers the required information level to easily find the best sellers. Considering numerous competitor firms, the opportunity of searching for a great variety of product models and a wide range of prices for a product, provided by different companies, not only make the product's life cycle shorter but also create a more fluid market (Baye et al. 2007).

2. Price skimming: This strategy divides customers into two price- and quality-sensitive groups. For performing this policy, firms set the initial price of new products so high that they achieve a noticeable profit by completely covering quality-sensitive customer demand. The prices then will be lowered over time to both attract low willing-to-pay customers and provide

firms the opportunity to survive in the competitive market, which leads to more revenue (Crettez, Hayek, and Zaccour 2020).

3. Penetration pricing: Contrary to the previous strategy, penetration pricing uses the cheapest possible price as the initial price of new products. By doing so, companies can easily attract the majority of market demand just on their arrival to the market, which helps them to achieve a considerable profit. When enlarging market share for a new brand is the main objective of companies, this policy would be their first priority (AlJazzazen 2019).

4. Customer perceived value: This strategy is based on equity theory; that is, equality between the company's and customer outcome/input ratio. If customers feel this comparison is fair and equitable, they will be satisfied and buy the products. So, to achieve a considerable market share and profit, the companies have to create a sense of equity in customers by delicately setting the prices (Qingyi and Ling 2019).

2.3 DEMAND FUNCTIONS

As mentioned by Duan and Ventura (2020), pricing policies can directly affect customer behavior and consequently influence both market demand and sellers' profit. In this competitive market, determining the best demand functions to help firms adjust their selling prices for not losing the customers is so crucial that numerous demand models are presented in the literature to formulate the relationship between a product's price and consumer demand. A comprehensive review of the most common ones is proposed in this section.

2.3.1 SINGLE-FIRM DEMAND FUNCTIONS

Table 2.1 summarizes the single-firm deterministic and stochastic price-dependent demand models (Huang 2013). Some of the most commonly used notations are explained as follows: ε: price-independent random variable, p: selling price, a: maximum purchase potential, b: price elasticity, d: item's demand, r: reference price, t: time period, and ζ: demand elasticity.

TABLE 2.1
Single-Firm Price-Dependent Demand Models

		Price-dependent models	Parameters
Linear Models	Deterministic	$d(p) = a - bp$	$a, b > 0$
		$d(p) = a(t) - bp(t)$	$a(t), b > 0, t$ $\in [0, T]$
		$d(p) = a(t) - b(t)p(t)$	$t \in [0, T]$
		$d(p, r) = a - \delta p(t) - \gamma \ (p(t) - r)$	$a, \delta, \gamma > 0$
	Stochastic	$d(p, \varepsilon) = d(p) + \varepsilon$	

		Price-dependent models	Parameters
Power /Isoelastic/ Constant-Elasticity Models	Deterministic	$d(p) = ap^{-b}$	$a > 0, b > 1$
		$d(p) = (ap + b)^{-\zeta}$	$a, b > 0, \zeta > 1$
		$d(p) = (a - bp)^{\zeta}$	$a, b, \zeta > 0, \zeta \in (-\infty, -1) \in (0, \infty)$
		$d(p) = a - bp^{\gamma}$	$a, b > 0, \gamma \geq 1$
		$d(p) = (a - bp^{\beta})^{\zeta}$	$a, b > 0, \beta \geq 1, \zeta \leq 1$
		$d(p) = a / [a + bp^{\gamma}]$	$a, b, \gamma > 0$
	Stochastic	$d(p, \varepsilon) = a\varepsilon / (\varepsilon + p^{b})$	
Hybrid Model	Deterministic	$d(p) = \tau(a_1 - bp) + (1 - \tau)a_2 p^{-\gamma}$	$a_1, a_2 > 0, b > 0, \gamma > 1, 0 \leq \tau \leq 1$
Exponential Models	Deterministic	$d(p) = a \exp(-bp)$	$a > 0, b > 0$
		$d(p) = \exp(a - bp)$	$a > 0, b > 0$
		$d(p) = a - \exp(p)$	$a > 0, b > 0$
	Stochastic	$d(p, \varepsilon) = \exp(\varepsilon - bp)$	
Logarithmic Model	Deterministic	$d(p) = \ln[(a - bp)^{\gamma}]$	$a, b, \gamma > 0$
	Stochastic	$d(p, \varepsilon) = \log_a^{\varepsilon - bp}$	
Logit Models	Deterministic	$d(p) = a\exp(-bp) / [1 + \exp(-bp)]$	$a > 0, b > 0$
		$d(p) = C\exp(-a - bp) / [1 + \exp(-a - bp)]$	
		$d(p) = 1 / [1 + \exp[a + bp]]$	$C > 0, a < -2, b > 0$
	Stochastic	$d(p, \varepsilon) = a\exp(\varepsilon - bp) / (1 + \exp(\varepsilon - bp))$	
General Models	Stochastic	$d(p, \varepsilon) = d(p)\varepsilon$	
		$d(p, \varepsilon) = d_1(p)\varepsilon + d_2(p)$	
		$d(p, \varepsilon) = \varepsilon_1 d(p) + \varepsilon_2$	

2.3.2 Competition-Based Multi-Firm Demand Functions

Multi-player price-dependent demand functions are also proposed in Table 2.2.

2.3.3 Advantages and Disadvantages of Different Functions

Making a comparison between the application of the presented demand models has initially revealed that all price-dependent demand models follow the general demand rule; that is, the higher the price of a product, the lower the demand sellers will suffer for that product (Duan and Ventura 2020). Moreover, some of their

TABLE 2.2

Multi-Player Price-Dependent Demand Functions

Linear Model		

$$d_i(p) = a_i - b_i p_i + \delta_i \sum_{j \neq i} p_j$$

where:

$$a_i = \alpha / \left[\beta + (n-1)\gamma \right]$$
$$b_i = \left[\beta + (n-2)\gamma \right] / \left[(\beta + (n-1)\gamma)(\beta - \gamma) \right]$$
$$\delta_i = \gamma / \left[(\beta + (n-1)\gamma)(\beta - \gamma) \right]$$

$\alpha, \beta > 0$: the effect of product i's price on its demand
$\gamma > 0$: the effect of the other products' prices on product i's demand ($\beta > \gamma$)

$$d_i(x_i, P_i, P_j) = x_i - \beta p_i + \gamma p_{3-i}$$

where

$$\int_{-\infty}^{\infty} f_i(x_i) dx_i = 1$$

For only two retailers ($i=1,2$)

x_i: a random variable of the demand quantity

$f_i(x_i)$: Probability density distribution function of x_i

Attraction Models	With linear attraction function for each firm	$d_i(p) = \dfrac{M(\alpha_i - \beta_i p_i)}{(V_0 + \sum_{j=1}^{n}(\alpha_j - b_j p_j))}$	M: market size V_0: the attractiveness of no-purchase $\alpha_i, \beta_i, V_0 \geq 0$
	With multinomial logit attraction function	$d_i(p) = \dfrac{M(\exp(\alpha_i - \beta_i p_i))}{V_0 + \sum_{j=1}^{n}(\exp(\alpha_j - \beta_j p_j))}$	$\alpha_i > 0$: the product's value anticipated by consumers $\beta_i > 0$: price sensitivity
	With multiplicative competitive interaction attraction function	$d_i(p) = M(\alpha_i p_i^{-\beta_i})$ $/ (V_0 + \sum_{j=1}^{n}(\alpha_j p_j^{-\beta_j}))$	$\alpha_i > 0$: product i's quality $\beta_i > 1$: price sensitivity
Multiplicative (Cobb–Douglas) Models		$d_i(p) = a_i P_i^{-\beta_i} \left(\prod_{j \neq i} p_j^{\beta_{ij}} \right)$	$a_i > 0$, $\beta_i > 1$: absolute price elasticity of firm i's $\beta_{ij} \geq 0$; $(j = 1, 2, \ldots, n, j \neq i)$: cross elasticity of firm i's demand to other firms
Constant expenditure Models		$d_i(p) = \gamma P_i^{-1} g(p_i) / \sum_{j=1}^{n} g(p_j)$	$\gamma > 0$ $g(p_i)$: a positive, strictly decreasing function of p_i

advantages and drawbacks are also mentioned in the previous research works, which are elaborated upon hereafter.

Considering the linear model, the following main pros can be mentioned as the major reasons for its extensive application in the related literature. Initially, it is the simplest function that can be used as demand models (Oum 1989). It also provides explicit solutions that help decision-makers easily interpret them and make optimal decisions. However, since the actual relationship between price and demand is rarely linear, this function is not realistic (Mahootchi and Husseini 2015). Moreover, the premise of assuming a uniform function as the customers' willingness to pay (WTP) in this model makes the decision-makers determine a finite upper boundary for the price, which also seems unrealistic.

Regarding the power model, since it can be both attained from the Cobb–Douglas production function and easily converted to a linear function, it is the most preferable in formulating non-linear demand–price relationships in the market. However, the constant demand elasticity $(-\zeta)$—in all prices, regardless of the demand function's location—is the main disadvantage of it. Also, due to its formulation, as the product's price goes to zero, its demand approaches infinity. Thus, since there is a specific size for each market, which is not infinite, the infinity demand (and subsequently the power function) looks unrealistic. Meanwhile, the exceeding of the demand function from the market size occurs in the exponential function (with a continuous increasing trend) when the price is very low. Considering WTP, the power function also proposes a downward curve in p, which never reaches zero.

To compensate the main weaknesses of the linear and convex (exponential and power) demand functions; that is, ignoring actual market characteristics and overlooking the realistic modeling of the price sensitivity in different values, logit demand function—otherwise known as reversed S-curve or the customer behavior model—is also presented. As mentioned in Duan and Ventura (2020), the linear demand function assumes customers' WTP is distributed uniformly from the lower to the upper boundary of the price; it has been proven that the middle of this range covers most customers' WTP, and the other parts are less representative (Bodea and Ferguson 2014). So, such a bell-shaped function as logit's explains this feature more realistically than a linear, exponential, or power one, which adds to the attraction of this function to be used in the research. More importantly, determining a price range for products has been one of the main challenging requirements of linear and other convex functions. According to the formulation of the logit function, when the price goes to zero, demand approaches the market size, and it goes to zero when the price converges to infinity. Hence, by taking into account logit formulation as the demand function, researchers would not have any necessity of defining any lower and upper boundaries for the prices, which can dramatically increase this function's desirability. Besides the prementioned advantages, enjoying the opportunity of both estimating this function's parameter by regression methods and proposing the most realistic representation of decision-makers' mode-switching manners are other eye-catching features of it.

All in all, power and linear demand functions are the most popular ones in the related literature. The reason for choosing one of them over the other has not been clearly stated in the papers, and seemingly, authors use them randomly (Yaghin, Ghomi, and Torabi 2014). In a special case, (Huang 2013) mentioned that the number of echelons in a system can determine the best-applied function. That is, if the system is a single-echelon one, all functions can lead to similar results, and this would provide the researchers the opportunity of selecting them randomly. However, they must pay more attention to the function selection process in multi-echelon systems since different functions result in different conclusions. Nevertheless, when it comes to the logit function, Duan and Ventura (2020) stated that the range of price deviation is the main selection criteria. Linear or power curves can perfectly formulate the small variation in the price, but forecasting demand in an extremely large range of prices requires applying the continuous price adjustment, which is provided in the logit function. It is worth clarifying that in the case of a fixed-demand elasticity, using power function is the smartest choice.

2.4 PRICING MODEL APPLICATIONS

These models have been used in many different industries, such as manufacturing, retailing, e-commerce, and hospitality, some practical studies and mathematical optimization models of which are presented as follows. According to Kienzler and Kowalkowski (2017), more than half of the pricing research papers concentrate on goods, and this number is much fewer for "focused-on-service" articles (59% and 23%, respectively). It is worth mentioning that a recent research trend has revealed a shift from goods to services studies. "Good-related" papers are also classified into two different categories of business-to-business (B2B)—transactions between businesses—and business-to-customer (B2C)—direct transactions between businesses and customers. The latter forms 62% of related articles, while 48% of them cover B2B topics. As an independent research category, the number of B2C pricing studies, which mostly consider the demand side of the supply chain, are four times the number of B2B articles. Psychological, differential, product-line, and competitor-oriented pricing are the most common topics of B2C papers, while domestic and international market conditions are the main determining strategies for B2B studies.

2.4.1 INDUSTRIES AND EXAMPLES

To find the effects of pricing strategies in practical subjects, their applications in two real problems are assessed in this section.

Transportation demand management (TDM) is defined as a set of strategies used to change commuters' travel behavior to improve the transportation system efficiency by decreasing travel demand and traffic congestion or by shifting trips in time or space (www.actweb.org). Many different strategies, such as parking management, congestion pricing, employer-based transportation programs, and land use, to name but a few, are assumed as the main TDM components. Specifically, pricing plays a critical role in TDM in two direct and indirect ways, which are explained here (Ferguson 1990).

   Considering the indirect effect of pricing in TDM, its influence in parking management strategies is noticeable. Parking management involves a variety of techniques used for making the best use of parking facilities by both changing transportation modes to "multi-occupant vehicles" and decreasing parking demand. Park sharing, space allocation to different categories of users, and pricing are the most important methods that can be classified as parking management strategies. Generally, parking pricing diminishes the total number of "single-occupant vehicle" trips in the priced areas and increases the shared-car travels and public transportation in the very region. Variable parking rates, dynamic parking tax, parking cash-out, and unbundling parking costs are the main pricing strategies that help the parking management process (Litman 2016).
   Pricing strategies can also directly affect TDM in a variety of ways. Congestion pricing (CP), defined as a variety of disincentive methods used to shift congestion in space during peak hours, is the most noticeable strategy in traffic management. Different toll rates for various highway lanes, entire roadways, and freeway ramps, as well as cordon and area-wide charges, are the main CP strategies (United States Department of Transportation—Federal Highway Administration). All in all, commuter traffic demand is insensitive to gasoline price since fuel makes up only a small part of driving costs. Conversely, it is highly sensitive to parking management strategies and congestion pricing, so it can be mentioned as the most reliable traffic management strategy (Steiner 1992).
   The application of pricing in electricity market management is another practical instance, which is elaborated upon here. Demand response is a program that focuses on end-customers to change their normal consumption patterns by varying the price of electricity or providing them some incentives or disincentives (U.S. Department of Energy). The beneficiaries of running this program are customers, system operators, and the environment. It reduces both the capital and operation costs of the companies— as the main consumers—enhances the grid stability, decreases the investment in improving the power grid, shifts the peak demand, and reduces carbon emission by curtailing electricity consumption and increasing renewable energy usage. This program is classified into two categories of price- or incentive (event)-driven types (Hussain and Gao 2018).
   Price-driven demand response (PDDR) is defined as a set of financial motivations or punishments to increase grid flexibility. Such different pricing programs as real-time pricing (RTP), critical peak pricing (CPP), time of use (TOU), and peak-time rebate (PTR), are its major strategies. In RTP pricing strategy, the price of electricity changes continuously due to real-time supply and demand (Widergren et al. 2012). In a TOU program, a static pricing portfolio for both peak and off-peak seasons is proposed to encourage customers to change their consumption behavior (Di Cosmo, Lyons, and Nolan 2014). PTR provides a reward (as a rebate in their bill), usually for residents or small businesses, to shift or decrease their energy use during peak events (Hussain and Gao 2018). In a CPP program, consumers receive financial motivation in extreme peak periods to either shift their consumption from peak hours or reduce it during that period. This strategy is normally used in warm summers or cold winters during which a high demand for power causes serious problems for the system.

Regarding the differences between CPP and TOU programs, CPP's value is much higher than TOU's pricing rate. Moreover, the former's usage period is limited to only a few days—when the generating utilities cannot cover total energy—while the latter provides a fixed tariff for most days of the year (Mohagheghi et al. 2010).

2.5 MATHEMATICAL MODELS

Two instances of mathematical formulations used to model related problems in the literature are presented hereafter.

2.5.1 FIRST MODEL

Burwell et al. (1991) presented an optimization model to maximize retailers' profit by determining lot size, pricing, and order-level variables; allowing shortages; and offering discounts. Demand is considered a price-dependent (monotonically decreasing) function in their study.

Terminology and Notations

Parameters		Decision variables	
C_0	Cost of ordering	P	Price
C_s	Cost of shortage	Q	Lot size
C_m	Variable cost of marketing	S	Order level
r	Unit cost of transportation		
v	The unit cost charged to the retailer		
$D=D(p)$	Demand for the product		

Model Development

$$\text{Maximize}: \pi\left(P,Q,S\right) = PD - vD - C_m D - \left(C_0 D/Q + C_s \left(Q - S\right)^2/2Q + rvS^2/2Q\right) \quad (2.1)$$

Solution Approach

The first and second derivatives of the objective function (Eq. (2.1)) with respect to P have shown that total profit is a concave function of P considering fixed values for both Q and S. So, the optimal value of P, $P^*\left(Q\right)$, is achieved by solving Eq. (2.2):

$$\partial\pi/\partial p = D + [P - (C_0/Q) - v - C_m]D' = 0 \quad (2.2)$$

Likely, a similar procedure has revealed that the objective is also a concave function of S for given values of p and Q; hence, Eq. (2.3) can be used to determine $S^*\left(Q\right)$:

$$C_s\left(Q - S\right) - rvS = \left(C_s\left(Q - S\right) - rvS\right)/Q = 0 \quad (2.3)$$

Finally, to find the optimal value of Q, P, and S, initially the profit function is derived by assuming Q as its only decision variable: $\pi_1\left(Q\right) = \pi\left(P^*\left(Q\right), Q, S^*\left(Q\right)\right)$; then, one iterative solution procedure (shown in Figure 2.2) has been used:

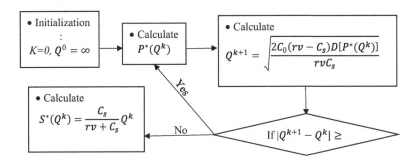

FIGURE 2.2 The solution approach procedure of the first model.

2.5.2 SECOND MODEL

Duan and Ventura (2020) used a logit price–dependent demand function (Eq. (2.9)) to find the optimal values of pricing, supplier selection, order quantities, and order frequencies with the aim of total profit maximization. The detailed description is presented as follows:

Terminology and Notations

Parameters

K_i Setup cost of ith supplier

c_i Unit purchasing cost from i^{th} supplier

F_i Total capacity of ith supplier

q_i Product quality level of ith supplier

q_a The average product quality which is required

h Inventory holding cost (unit per time unit)

m Maximum orders numbers in a cycle

D Product demand

Sets

r Set of suppliers, indexed by i

Decision variables

P Unit price

J_i Number of orders from i^{th} supplier in a cycle

Q_i Quantity of orders from i^{th} supplier

Q' Overall order quantity in a cycle

T Length of a cycle

T_i Time to consume an order from i^{th} supplier

Model Development

$$\text{Maximize } Z = DP - \frac{D}{Q'}\sum_{i=1}^{r} J_i K_i - \frac{h\sum_{i=1}^{r} J_i Q_i^{2}}{2Q'} - D\frac{\sum_{i=1}^{r} J_i c_i Q_i}{Q'} \tag{2.4}$$

$S.t.$

$$Q' = \sum_{i=1}^{r} J_i Q_i . \tag{2.5}$$

$$DQ_i J_i / Q' \leq F_i \qquad\qquad \forall i \tag{2.6}$$

$$\sum_{i=1}^{r} J_i q_i Q_i \geq q_a \sum_{i=1}^{r} J_i Q_i . \tag{2.7}$$

$$\sum_{i=1}^{r} J_i \leq m \tag{2.8}$$

$$D(p) = C(e^{-(a+bp)} / 1 + e^{-(a+bp)}) \qquad\qquad C > 0, a < -2, b > 0 \tag{2.9}$$

$$J_i \geq 0, integer \qquad\qquad \forall i \tag{2.10}$$

$$Q_i \geq 0, D \geq 0, P \geq 0 \qquad\qquad \forall i \tag{2.11}$$

With the goal of maximizing total profit, the constraints of not exceeding both the suppliers' capacity and the maximum number of orders in a cycle as well as meeting product quality requirements are also formulated.

Solution Approach

A two-stage piecewise-linear approximation algorithm is applied to find near-optimal solutions. To do so, the authors have used Lundell's (2009) proposed method to determine their breakpoint selection policy.

2.6 RESEARCH TRENDS

Determining the main trends in pricing policy has a considerable effect on improving researchers' creativity, which leads to the formation of novel values and contributions. Thus, with the goal of presenting the recent research trend, in this section, most related studies are classified and proposed as different categories. These groups and their articles are presented as follows.

2.6.1 INVENTORY AND PRICING

For the first time, Whitin (1955) considered economic order quantity models in pricing decisions with the goal of determining price and inventory level simultaneously. By assuming a linear price-dependent demand function, he applied a sequential algorithm to find the optimal price by receiving the inventory level, which was calculated in the previous stage. Hartwig et al. (2015) investigated the influence of strategic inventories on performing a supply chain in a two-period problem with a price-dependent demand function. This paper has shown that it improves the efficiency of the supply chain by narrowing the supplier's power in the market.

Moreover, Güler, Bilgiç, and Güllü (2015) evaluated the effects of customers' reference price and the product's retail price in the profit of a supply chain. Simultaneous optimization of pricing and inventory decisions was the main objective of this article, achieved by developing a multi-period stochastic price-dependent demand model. Taleizadeh and Noori-Daryan (2016) used a Stackelberg game—Nash equilibrium concept—in a two-echelon supply chain to find pricing decisions and ordering quantities between each two presented layers. Covering of all customer demands (without any shortage) and minimizing supply chain cost were the major goals of this article. In a similar study, Heuvel and Wagelmans (2006) formulated a deterministic, multi-period demand-dependent model in an economic lot size model to make pricing and ordering decisions optimally.

Díaz-Mateus et al. (2018) proposed a constrained multinomial logit function to find the demand level in a three-layer supply chain and take into account both customers' willingness to pay function and their economic as well as social differences. Also, Wu et al. (2017) studied product expiration dates and their life cycles; first, he calculated items' deterioration rate and then explored their freshness-dependent prices. In a similar study, Jadidi, Jaber, and Zolfaghari (2017) studied optimizing such decisions as pricing, inventory, transportation capacity, and transportation cost to maximize the profit of both players of the supply chain and customers in a single-period, single-product, supplier–buyer chain.

2.6.2 FACILITY LOCATION AND PRICING DECISIONS

Mahootchi and Husseini (2015) developed a multi-period, multi-product, multi-layer network to satisfy customers' price-dependent demand in a supply chain. In this study, not only were designing and redesigning decisions, pricing, and capacity planning investigated, but budget and capacity limitations were also formulated to make the study closer to reality. Afterward, Fattahi, Govindan, and Keyvanshokooh (2017) considered a multi-period supply chain and classified the problem's decisions into two categories of tactical (like retailing) and strategic (such as redesigning of the supply chain) where demand is a stochastic price-dependent function.

Making pricing, location, transportation, and production decisions in a multi-leader, multi-follower game, where all players are independent supply chains and a multi-product supply chain versus supply chain competition holds, is studied by Saghaeeian and Ramezanian (2018). In this competition, all new and existing supply chains compete to satisfy more demand, which behaves like a stochastic linear-dependent function. Similarly, Rezapour et al. (2015) considered a supply chain, which is responsible for producing and distributing new products, and a new rival supply chain, which can provide not only new products but also reproduced ones to the market. Consequently, two internal and external competitions: for supplying products in the new supply chain and between two chains for covering more price-dependent demand are considered.

Managing returned products, which is a function of the return price, in a closed-loop supply chain with different product qualities has been surveyed by Farshbaf-Geranmayeh, Taheri-Moghadam, and Torabi (2020). Evaluating the profitability of providing costly incentives to increase the number of returned products in the case of not reaching a specific returned products' threshold is one of this article's challenges. Likewise, simultaneous optimization of reverse and forward flows of a multi-stage, closed-loop supply chain was studied by Atabaki, Khamseh, and Mohammadi (2019). In this mixed-integer linear programming network design model, customer zones have price-dependent demand. Such realistic constraints as the capacity and number of facilities have been added to the mathematical model to turn it to an accurate formulation of real problems.

2.6.3 SUPPLIER SELECTION

Making joint procurement and selling price decisions in a newsvendor model by considering different suppliers and "the total cost before selling season" has been studied by X. Hu and Su (2017). It has revealed that the more preseason cost the seller has to pay, the less optimal order quantity he or she will have, the higher optimal price customers will suffer, and the less profit the newsvendor will achieve. Moreover, it has been proven that first, separate procurement and selling policies results the weaker performance compared to the integrated strategies, and second, both total profit and purchasing time rise as the result of increasing the number of suppliers.

Noori-Daryan, Taleizadeh, and Jolai (2017) also provided an investigation for a single-product supply chain including multi-national suppliers, whose production capacity is limited, and a retailer; the retailer suffers from uncertainty in the response time of his or her order, and demand is both price and delivery lead time dependent. Moreover, selecting suppliers from among the potential ones, determining order quantities as well as the optimal number of orders, and optimizing the selling price for one type of product in a single-retailer multi-supplier supply chain has been studied by Adeinat and Ventura (2015). By considering Karush–Kuhn–Tucker conditions, they also have shown the effects of suppliers' capacity in the sourcing process under capacity and quality constraints. Combining dynamic pricing and the supplier selection problem in a multi-period, single-product supply chain with supply and demand uncertainties and a price-dependent demand has been addressed by Feng and Shi (2012). Determining the influence of multiple sourcing on total expected profit and contrasting it with dynamic pricing effects was another contribution of this article. An extension to their model was presented by Duan and Ventura (2020). They used a logit price-dependent demand function to find an optimal solution for supplier selection, inventory decisions, and pricing problems simultaneously. The related literature is summarized in Table 2.3.

According to Table 2.3, three main topics can be mentioned for future research. First, most of the pricing articles consider linear price-sensitive demand function since it is the simplest function and can lead the authors to the closed-form solution easily. However, as it is clarified in Section 2.3.3, there are some other functions that can model the actual behavior of demand more precisely. So, investigating different types of demand functions and using them in future pricing articles make related research studies more practical. Second, articles that addressed inventory variables or transportation flows usually considered one supplier in the problem, whereas it sounds more realistic if researchers model multi-supplier cases in their papers and formulate them as supplier selection problems as well. Third, taking into account disruption risk in the supply chain design and redesign with price-dependent demand and making location and relocation decisions in multi-leader multi-follower games with demand uncertainty are other possible contributions for future studies.

TABLE 2.3

Pricing in the Supply Chain–Related Literature

Reference	SC structure	Mathematical model			Solution approach
		Decision variables	Objective function	Demand function	
Feng and Shi (2012)	Multiple potential suppliers, Single retailer	Dynamic pricing, Supplier selection, Ordering	Expected profit maximization	Linear, Stochastic	Convex analysis
Hsieh, Chang, and Wu (2014)	Multiple manufacturers, Single retailer	Competitive pricing, Ordering	Profit maximization	Linear, Stochastic	A heuristic procedure
Rezapour et al. (2015)	Supply chain versus supply chain competition	Competitive pricing, Inventory, Facility location	Profit maximization	Linear, Deterministic	A modified projection solution technique
Mahootchi and Husseini (2015)	Potential production plants, Warehouses, Customer zones	Pricing, Network design, Network redesigning, Capacity planning	Supply chain net income maximization	Linear, Deterministic	CPLEX software, A heuristic approach
Hartwig et al. (2015)	Single supplier, Single buyer	Pricing, Inventory	Profit maximization	Linear, Deterministic	Backward induction
Güler, Bilgiç, and Güllü (2015)	Single vendor, Single buyer	Pricing, Inventory	Profit maximization	Linear, Stochastic	Stochastic dynamic programing (convex analysis)
Adeinat and Ventura (2015)	Multiple potential suppliers, Single retailer	Pricing, Supplier selection, Ordering	The retailer's profit maximization	Power, Deterministic	A heuristic approach
Taleizadeh and Noori-Daryan (2016)	Single supplier, Single producer, Multiple retailers	Supplier's and producer's pricing, The number of shipments received by the supplier and producer	Supply chain total cost optimization	Linear, Deterministic	Stackelberg game

(Continued)

TABLE 2.3

Pricing in the Supply Chain–Related Literature

Reference	SC structure	Mathematical model			Solution approach
		Decision variables	Objective function	Demand function	
Jadidi, Jaber, and Zolfaghari (2017)	Single supplier Single buyer	Pricing Inventory	Profit maximization	Linear Stochastic	A proposed Sudoku (brute-force) algorithm
Wang, Chen, and Chen (2017)	Single supplier Single firm	Selling and wholesale pricing Order quantity	Retailer's profit maximization	Linear Stochastic	An exact method
Wu et al. (2017)	Single supplier Single retailer	pricing Order quantity	Profit maximization	Exponential Deterministic	Differential equations
Díaz-Mateus et al. (2018)	Single supplier Single vendor	pricing Inventory coordination	Profit maximization	Constrained multinomial logit Stochastic	A metaheuristic approach based on particle swarm optimization
Fattahi, Govindan, and Keyvanshokooh (2017)	Production plants Warehouses Customer zones	Pricing Location Capacity Production Transportation Inventory decisions	Supply chain net income and unspent budget maximization	Linear Stochastic	Benders decomposition algorithm
Noori-Daryan, Taleizadeh, and Jolai (2017)	Multiple potential suppliers Single retailer	Pricing Promised lead time Ordering Supplier selection	Supply chain profit maximization	Linear Deterministic	Three game-theoretic approaches
X. Hu and Su (2017)	Multiple potential suppliers Single retailer	Pricing Supplier selection Purchasing time Order quantity	The newsvendor's expected profit maximization	Linear Stochastic	A heuristic approach

Reference	SC structure	Mathematical model			Solution approach
		Decision variables	Objective function	Demand function	
B. Hu, Qu, and Meng (2018)	Single supplier Single retailer	Product retail price Option price Option exercise price Order quantity	Profit maximization	Linear Stochastic	A sequential procedure
Saghaeeian and Ramezanian (2018)	Supply chain versus supply chain competition	Competitive pricing Location Transportation Production decisions	Leader's and follower's profit maximization	Linear Deterministic	A hybrid genetic algorithm
Duan and Ventura (2020)	Multiple potential suppliers Single retailer	pricing Ordering Supplier selection	Profit maximization	Logit Deterministic	A two-stage piecewise linear approximation algorithm
Farshbaf-Gerannayeh, Taheri-Moghadam, and Torabi (2020)	A closed-loop supply chain, including: Single production plants Outsourcing centers Distribution centers Customer zones Collection centers	Pricing Inventory network design and its forward and reverse flows	Expected profit maximization	Linear Stochastic	Hybrid genetic algorithm and simulated annealing GAMS software
Atabaki, Khamseh, and Mohammadi (2019)	A closed-loop network, including: Single supplier Production plants Distribution centers Redistributors Hybrid distributor and redistributor Disassembly and disposal centers	Pricing Location Transportation flows Allocation	Profit maximization	Discrete	GAMS/CPLEX software A metaheuristic approach

REFERENCES

Adeinat, Hamza, and Jos A. Ventura. 2015. Determining the retailer's replenishment policy considering multiple capacitated suppliers and price-sensitive demand. *European Journal of Operational Research* 247, no. 1: 83–92.

AlJazzazen, Sahoum Ali. 2019. New product pricing strategy: Skimming Vs. Penetration. In *Proceedings of FIKUSZ Symposium for Young Researchers*, 1–9. Budapest, Hungary: Óbuda University Keleti Károly Faculty of Economics.

Atabaki, Mohammad Saeid, Alireza Arshadi Khamseh, and Mohammad Mohammadi. 2019. A Priority-based firefly algorithm for network design of a closed-loop supply chain with price-sensitive demand. *Computers & Industrial Engineering* 135: 814–837.

Baker, Walter L., Michael V. Marn, and Craig C. Zawada. 2010. *The Price Advantage*. Vol. 535. Hoboken, NJ: John Wiley & Sons.

Baye, Michael R., J. Rupert, J. Gatti, Paul Kattuman, and John Morgan. 2007. A dashboard for online pricing. *California Management Review* 50, no. 1: 202–216.

Blythe, Jim. 2006. *Essentials of Marketing Communications*. Harlow: Pearson Education.

Bodea, Tudor, and Mark Ferguson. 2014. *Segmentation, Revenue Management and Pricing Analytics*. New York: Routledge.

Brassington, Frances, and Stephen Pettitt. 2006. *Principles of Marketing*. Harlow: Pearson Education.

Burwell, Timothy H., Dinesh S. Dave, Kathy E. Fitzpatrick, and Melvin R. Roy. 1991. An inventory model with planned shortages and price dependent demand. *Decision Sciences* 22, no. 5: 1187–1191.

Cannon, Hugh M., and Fred W. Morgan. 1990. A strategic pricing framework. *Journal of Services Marketing* 4, no. 2:19–30.

Cavallo, Alberto. 2018. *More Amazon Effects: Online Competition and Pricing Behaviors*. Cambridge, MA: National Bureau of Economic Research.

Cosmo, Valeria Di, Sean Lyons, and Anne Nolan. 2014. Estimating the impact of time-of-use pricing on irish electricity demand. *The Energy Journal* 35, no. 2.

Cressman, George E. 2012. Value-based pricing: A state-of-the-art review. In *Handbook of Business-to-Business Marketing*. Cheltenham: Edward Elgar Publishing.

Crettez, Bertrand, Naila Hayek, and Georges Zaccour. 2020. Existence and characterization of optimal dynamic pricing strategies with reference-price effects. *Central European Journal of Operations Research* 28, no. 2: 441–459.

Díaz-Mateus, Yeison, Bibiana Forero, Héctor López-Ospina, and Gabriel Zambrano-Rey. 2018. Pricing and lot sizing optimization in a two-echelon supply chain with a constrained logit demand function *International Journal of Industrial Engineering Computations* 9, no. 2: 205–220.

Duan, Lisha, and Jose A. Ventura. 2020. Technical note : A joint pricing, supplier selection, and inventory replenishment model using the logit demand function. *Decision Sciences*: 1–23.

Elmaghraby, Wedad, and Pınar Keskinocak. 2003. Dynamic pricing in the presence of inventory considerations: Research overview, current practices, and future directions. *Management Science* 49, no. 10: 1287–1309.

Faith, Dudu Oritsematosan, and Agwu M. Edwin. 2014. A review of the effect of pricing strategies on the purchase of consumer goods. *International Journal of Research in Management, Science & Technology* 2, no. 2: 88–102.

Farese, Lois Schneider, Grady Kimbrell, and Carl A. Woloszyk. 2001. *Marketing Essentials*. New York: Glencoe/McGraw-Hill.

Farshbaf-Geranmayeh, Amir, Alireza Taheri-Moghadam, and S. Ali Torabi. 2020. Closed loop supply chain network design under uncertain price-sensitive demand and return. *Information Systems and Operational Research*: 1–29.

Fattahi, Mohammad, Kannan Govindan, and Esmaeil Keyvanshokooh. 2017. A multi-stage stochastic program for supply chain network redesign problem with price-dependent uncertain demands. *Computers and Operations Research* 100: 314–332.

Feng, Qi, and Ruixia Shi. 2012. Sourcing from multiple suppliers for price-dependent demands. *Production and Operations Management* 21, no. 3: 547–563.

Ferguson, Erik. 1990. Transportation demand management planning, development, and implementation. *Journal of the American Planning Association* 56, no. 4: 442–456.

Gorodnichenko, Yuriy, and Oleksandr Talavera. 2017. Price setting in online markets: Basic facts, international comparisons, and cross-border integration. *American Economic Review* 107, no. 1: 249–282.

Güler, M. Güray, Taner Bilgiç, and Refik Güllü. 2015. Joint pricing and inventory control for additive demand models with reference effects. *Annals of Operations Research* 226, no. 1: 255–276.

Hartwig, Robin, Karl Inderfurth, Abdolkarim Sadrieh, and Guido Voigt. 2015. Strategic inventory and supply chain behavior. *Production and Operations Management* 24, no. 8: 1329–1345.

Heuvel, Wilco Van Den, and Albert P. M. Wagelmans. 2006. A polynomial time algorithm for a deterministic joint pricing and inventory model. *European Journal of Operational Research* 170: 463–480.

Hemmati, M., Fatemi Ghomi, S. M. T., and Sajadieh, M. S. 2017. Vendor managed inventory with consignment stock for supply chain with stock-and price-dependent demand. *International Journal of Production Research* 55, no. 18: 5225–5242.

Hinterhuber, Andreas, and Marco Bertini. 2011. Profiting when customers choose value over price. *Business Strategy Review* 22, no. 1: 46–49.

Hsieh, Chung-chi, Yih-long Chang, and Cheng-han Wu. 2014. Competitive pricing and ordering decisions in a multiple-channel supply chain. *International Journal of Production Economics* 154: 156–165.

Hu, Benyong, Jiali Qu, and Chao Meng. 2018. Supply chain coordination under option contracts with joint pricing under price-dependent demand. *International Journal of Production Economics* 205: 74–86.

Hu, Xiangling, and Ping Su. 2017. The newsvendor's joint procurement and pricing problem under price-sensitive stochastic demand and purchase price uncertainty. *Omega* 79: 81–90.

Huang, Jian. 2013. Demand functions in decision modeling: A comprehensive survey and research directions. *Decision Sciences* 44, no. 3: 557–609.

Hussain, Muhammad, and Yan Gao. 2018. A review of demand response in an efficient smart grid environment. *The Electricity Journal* 31, no. 5: 55–63.

Ingenbleek, Paul, Marion Debruyne, Ruud T. Frambach, and Theo M. M. Verhallen. 2003. Successful new product pricing practices: A contingency approach. *Marketing Letters* 14, no. 4: 289–305.

Jadidi, Omid, Mohamad Y. Jaber, and Saeed Zolfaghari. 2017. Joint pricing and inventory problem with price dependent stochastic demand and price discounts. *Computers & Industrial Engineering* 114: 45–53.

Jampala, Rajesh C. 2016. Pricing strategy. In *Strategic Marketing Management in Asia*, 383–402. Bingley, UK: Emerald Group Publishing Limited.

Kienzler, Mario, and Christian Kowalkowski. 2017. Pricing strategy: A review of 22 years of marketing research. *Journal of Business Research* 78: 101–110.

Kotler, Philip, and Gary Armstrong. 2010. *Principles of Marketing.* Upper Saddle River, NJ: Pearson Education.

LaPlaca, Peter J. 1997. Contributions to marketing theory and practice from industrial marketing management. *Journal of Business Research* 38, no. 3: 179–198.

Lichtenstein, Donald R., Nancy M. Ridgway, and Richard G. Netemeyer. 1993. Price percep-
   tions and consumer shopping behavior: A field study. *Journal of Marketing Research*
   30, no. 2: 234–245.
Lilien, Gary L., Philip Kotler, and K. Sridhar Moorthy. 1992. *Marketing Models*. Englewood
   Cliffs, NJ: Prentice Hall.
Litman, Todd. 2016. *Parking Management: Strategies, Evaluation and Planning*. Victoria,
   BC: Victoria Transport Policy Institute.
Lundell, Andreas. 2009. *Transformation Techniques for Signomial Functions in Global
   Optimization*. Finland: Åbo Akademi University Åbo.
Mahootchi, M., M. Fattahi, and S. M. Moattar Husseini. 2015. Integrated strategic and tactical
   supply chain planning with price-sensitive demands. *Annals of Operations Research*
   242, no. 2: 423–456.
Marn, Michael V., and Robert L. Rosiello. 1992. Managing price, gaining profit. *McKinsey
   Quarterly* 18.
Mohagheghi, Salman, James Stoupis, Zhenyuan Wang, Zhao Li, and Hormoz Kazemzadeh.
   2010. Demand response architecture: Integration into the distribution management sys-
   tem. 2010 First IEEE International Conference on Smart Grid Communications, IEEE,
   Gaithersburg 501–506.
Monroe, Kent B. 2002. *Pricing: Making Profitable Decisions*. 3rd Ed. New York: McGraw-
   Hill Book Company.
Noori-Daryan, Mahsa, Ata Allah Taleizadeh, and Fariborz Jolai. 2017. Analyzing pricing,
   promised delivery lead time, supplier-selection, and ordering decisions of a multi-
   national supply chain under uncertain environment. *International Journal of Production
   Economics* 209: 236–248.
Oum, Tee Hoon. 1989. Alternative demand models and their elasticity estimates. *Journal of
   Transport Economics and Policy*: 163–187.
Qingyi, Wang, and Xu Ling. 2019. Dynamic pricing model of retailer with consideration of
   customer perceived value. *Logistics Technology* 4: 7.
Ratchford, Brian T. 2009. Online pricing : Review and directions for research. *Journal of
   Interactive Marketing* 23, no. 1: 82–90.
Rezapour, Shabnam, Reza Zanjirani, Behnam Fahimnia, and Kannan Govindan. 2015.
   Competitive closed-loop supply chain network design with price-dependent demands.
   Journal of Cleaner Production 93: 251–272.
Saghaeeian, Amin, and Reza Ramezanian. 2018. An efficient hybrid genetic algorithm for
   multi-product competitive supply chain network design with price-dependent demand.
   Applied Soft Computing Journal 71: 872–893.
Sajadieh, Mohsen S., and Mohammad R. Akbari Jokar. 2009. Optimizing shipment, order-
   ing and pricing policies in a two-stage supply chain with price-sensitive demand.
   Transportation Research Part E: Logistics and Transportation Review 45, no. 4:
   564–571.
Seyedhosseini, Seyed Mohammad, Seyyed-Mahdi Hosseini-Motlagh, Maryam Johari, and
   Mona Jazinaninejad. 2019. Social price-sensitivity of demand for competitive supply
   chain coordination. *Computers & Industrial Engineering* 135: 1103–1126.
Steiner, Ruth L. 1992. Lessons for transportation demand management from utility industry
   demand-side management (Abridgment). *Transportation Research Record* 1346.
Taleizadeh, Ata Allah, and Mahsa Noori-Daryan. 2016. Pricing, manufacturing and inventory
   policies for raw material in a three-level supply chain. *International Journal of Systems
   Science* 47, no. 4: 919–931.
Wang, Chong, Jing Chen, and Xu Chen. 2017. Pricing and order decisions with option
   contracts in the presence of customer returns. *International Journal of Production
   Economics* 193: 422–436.
Whitin, T. M. 1955. Inventory control and price theory. *Management Science* 2, no. 1: 61–680.

Widergren, Steve, Cristina Marinovici, Teri Berliner, and Alan Graves. 2012. Real-time pricing demand response in operations. In *2012 IEEE Power and Energy Society General Meeting*, 1–5. San Diego: IEEE.

Wu, Jiang, Chun-tao Chang, Jinn-tsair Teng, and Kuei-kuei Lai. 2017. Optimal order quantity and selling price over a product life cycle with deterioration rate linked to expiration date. *International Journal of Production Economics* 193: 343–351.

Yaghin, R. Ghasemy, S. M. T. Fatemi Ghomi, and S. A. Torabi. 2014. Enhanced joint pricing and lotsizing problem in a two- echelon supply chain with logit demand function. *International Journal of Production Research* 52, no. 17: 4967–4983.

3 On-Shelf Inventory

Subrata Saha and Peter Nielsen
Department of Materials and Production,
Aalborg University, Denmark

CONTENTS

3.1 INTRODUCTION: BACKGROUND AND DRIVING FORCES

Display inventory is one of the most important aspects of retailing to stimulate demand and improve customer awareness. According to Levin (1972), "at times, the presence of inventory has a motivational effect on people around it. It is a common belief that large piles of goods displayed in a departmental store lead the customers to buy more." An empirical investigation by Desmet and Renaudin (1998) also established that shelf-space elasticity remains positive for various product categories, such as toys, consumer electronics goods, vegetables, clothes, furniture, books, sports equipment, musical instruments, and baked goods. Balakrishnan, Pangburn, and Stavrulaki (2004) state that the presence of inventories can stimulate demand for several reasons, such as "increasing product visibility, kindling latent demand, signaling a popular product, or providing consumers an assurance of future availability." This demand-stimulating influence of display inventory, sometimes referred to as the "billboard effect," is explained by reasons such as increased visibility, perception of the product's popularity, and consumer awareness. It is common to find glamorous and scientifically orientated displays of products in stores, referred to as "psychic stock" by retailers. Urban (2005) stated, "Operations management literature has recognized this motivating effect of inventory on product demand function, and models have been developed that incorporate this relationship." Empirical evidence from both marketing and operations management literature has demonstrated that demand is inventory-dependent in supermarkets, home appliance stores, and magazine retailers, among others (Koschat 2008). It motivates researchers as well as practitioners to explore the effect of variability of the inventory-level-dependent demand rate from different perspectives. Extensive literature in this stream of research exists (see, e.g., Silver and Meal 1969; Ritchie and Tsado 1985; Padmanabhan and Vrat 1995; Goyal and Chang 2009; Panda, Saha, and

Basu 2009; Yang, Teng, and Chern 2010; Teng et al. 2011; Saha, Das and Basu 2012; Yang 2014; Moon et al. 2017; Pando, San-José, and Sicilia 2020).

In this chapter, we primarily select representative publications on display-inventory-dependent demand based on the following three criteria, which are listed in order of their priority. First, we highlight the influences of some key factors associated with decision-making in these circumstances. Second, we emphasize the categorization of the demand functions when the demands of consumers are sensitive to retailers' decisions regarding display inventory. A mathematical model is used to demonstrate the computational complexity associated with the decision-making process. Finally, we summarize our work and discuss how the existing demand models can help us analyze a variety of problems and their limitations. We conclude this chapter with a discussion of potential research directions.

3.2 SOME KEY FACTORS ASSOCIATED WITH ON-SHELF INVENTORY

In-store inventory management continues to be a key issue in retail operations that affects not only the performance of the downstream firm—the retailer—but also members involved in the overall supply chain. It is well documented that modern retailers have increasingly benefited from the use of information technology for tracking inventory and technologies to improve operational efficiency. It is difficult to ignore the "billboard effect" in today's retail practice to obtain a robust replenishment decision. However, display inventories have both good and bad aspects, and a higher display inventory level sometimes discourages customers because it may indicate that the product may be unpopular or of low quality or suggest future availability at lower prices. Researchers sometimes attribute this phenomenon to the perception of exclusiveness, fear of stock-outs, or bandwagon effects. This damaging relationship between display inventory level and demand, which is relevant for products such as exclusive automobiles or gift items and special wines, is often called the "scarcity effect" (Cachon, Gallino, and Olivares 2019).

Therefore, how to balance the billboard and the scarcity effects of display inventory is always a challenging issue for retail firms. Moreover, an appropriate allocation scheme for the replenishment quantity between a retail store's backroom and shelf space is always an issue. Retailers' strategic measures regarding demand function estimation can, in this regard, cause an adverse effect on not only themselves but also on consumers as well as all associates in the supply chain. Several factors exist affecting such a complex decision-making process—we present some of them in Figure 3.1.

Price: Consumer demand is affected by many factors, and price is one of the key influences. In the literature, several researchers study the joint influence of display stock and the retail price. Since demand decreases as time progresses due to the decrement of display stock level, it is therefore always challenging for retailers to decide whether they need to set a uniform price or adjust the price dynamically as time progresses. Consequently, the following issues are relevant. First, if the retailer increases the selling price, then customers may move to other stores. Second, if the retailer decreases the selling price, then, due to the sudden rise in demand, the retailer may face stock-out problems. Third, if the retailer fixes the

Deterioration	**Trade Credit**
• Instantaneous (Dye and Ouyang 2005) • Non-instantaneous (Zhang, Wang, Lu, and Tang 2015)	(Min, Zhou, and Zhao 2010)

	Single vs. Multiple items
No zero ending inventory measure • Positive end inventory (Shaikh et al. 2019) • Partial backlogging (Yang, Teng, and Chern 2010)	• Substitute products (Krommyda, Skouri, and Konstantaras 2015) • Complimentary products (Dehghanbaghi and Sajadieh 2017)

Holding cost	**Price**
• Linear (Chen, Chen, Keblis, and Li 2019) • Non-linear (Alfares 2007)	• Static (Saha and Goyal 2015) • Dynamic (Dey, Chatterjee, Saha, and Moon 2019)

FIGURE 3.1   Factors affecting replenishment decisions and the main sources addressing them.

price throughout, then the retailer may also lose a potential opportunity to skim profit at the beginning of the replenishment cycle or lose potential consumers during the end of the cycle.

Remark: Dynamic pricing decisions under stock-dependent demand, whether it is a continuous (Dye and Ouyang 2005; Panda, Saha, and Basu 2009) or discrete-time (San-José et al. 2019) framework, remain an important research issue. More analysis is necessary for this direction because the static pricing models (Maiti and Maiti 2005; Cárdenas-Barrón et al. 2020) sometimes fail to replicate the pragmatic practice during the entire replenishment cycle, especially for deteriorating or fashionable products.

Holding cost: Appropriate display of inventory by preserving its freshness level during the replenishment cycle always requires additional investment. Therefore, the holding cost for a retailer should be time varying. However, perhaps due to analytical tractability, replenishment decisions are made by assuming holding cost as constant in many studies (Yang 2014), although there exists a potential amount of articles where holding cost is assumed as a nonlinear (Chang 2004; Urban 2008; Alfares 2007; Pando, San-José, and Sicilia 2019) increasing/decreasing function of time or even as a function of replenishment quantity. Therefore, the following issues are relevant. First, if the holding cost increases, in particular for deteriorating items toward the end of their lifespan, there is a potential scope to study the effect of dynamic pricing under stock-dependent demand. Second, it is always challenging to rent a large space in a busy supermarket; consequently, the retailer may use a warehouse or backroom to keep additional replenishment not only to reduce holding costs but also to decrease the deterioration rate as well as the ordering cost.

Remark: In a single- or multi-item replenishment decision process, how to categorize products based on holding cost and integrating strategic pricing decisions is another avenue of research under the influence of stock-dependent demand. For stock-keeping unit managers, it is important to select products to be displayed, with optimal allocation in limited display space to minimize deterioration and holding cost.

Deterioration: Ghare and Schrader (1963) first captured the effect of product deterioration and proposed the corresponding governing differential equation. The deterioration of most physical goods is common, and in the literature, the product deterioration rate is generally classified into two categories—instantaneous (Bhunia et al. 2014) and non-instantaneous (Zhang et al. 2015)—based on the nature of the products. Some misapprehensions exist regarding the starting time of product deterioration for the latter type of products within the replenishment cycle, but it is relevant to study replenishment decisions for these categories of products. Some items deteriorate continuously as time progresses; however, the rate is sometimes proportional to the amount of display inventory (Bakker, Riezebos, and Teunter 2012; Önal, Yenipazarli, and Kundakcioglu 2016). Some products, such as milk or bakery items, have fixed lifetimes, after which they deteriorate completely. The problem becomes more challenging when the items are perishable, such as milk, vegetables, and dairy products. Therefore, deterioration increases the depletion rate of the items in stock. Since the demand is affected by the display inventory level, how to keep them fresh is a fundamental challenge faced by the retailer. Although for deteriorating items, shelf-life restrictions force the retailers to keep fewer items in stock, and they tend to keep high levels of inventory in anticipation of higher demand (Goyal, Levi, and Segev 2016). In this regard, investment in preservation technology to slow down deterioration also becomes relevant (Hsu, Wee, and Teng 2010). Therefore, the following issues are to be noted: first, to capture the practicality, researchers used complex functions such as Weibull or normal distribution functions and a linear time-dependent deterioration function, and that leads to a complex nonlinear optimization problem. The determination of closed-form expression is almost impossible under such scenarios. Therefore, particularly in the supply chain environment, how to model the deterioration effect is a challenge. Second, another challenge is how investment in preservation technology affects the pricing decision as well as waste minimization to ensure future sustainability.

Remark: Empirical studies are necessary to validate assumptions regarding the deterioration function used in literature and its influence on replenishment decisions under stock-dependent demand.

Non-zero end inventory measure: The replenishment decision by allowing shortages is one of the fundamental factors, not only under the stock-dependent-demand function but also in several others. However, under stock-dependent demand, two groups of literature exist: (1) models where shortages are allowed (Tiwari et al. 2017), and (2) a positive ending condition for the inventory level at the end of the replenishment cycle (Yang 2014; Pando, San-José, and Sicilia 2019). In the literature, various models are formulated by allowing storages by assuming a complete (Khan et al. 2020) or partial backlogging opportunity (Duan et al. 2012). To replicate the effect of consumers' waiting time, models are formulated by assuming the constant, time-dependent, or exponential distribution function as a backlog rate (Chang, Goyal, and Teng 2006; Maihami and Abadi 2012). However, the second group of literature is somehow more important under stock-dependent demand. The main question is if demand is stock dependent, then the retailer might not wait until the inventory becomes zero or negative; rather, it will be more relevant to explore the characteristics of optimal decision that can control both initial and non-zero end inventory levels (Chen et al. 2016).

Remark: In the age of information technology, it is always crucial to study whether the functional relations used for representing the backlogging rate can replicate the section of consumers' minds whose purchase decision is affected by display stock. Second, a replenishment decision model that can deliver the optimal triplet price-initial-and-end-inventory decision is more pragmatic under stock-dependent demand.

Trade credit: In the changing landscape of the business environment, it is almost impossible to ignore the effect of firms involved in facilitating credit, whether under the guise of trade credit, trade finance, working capital, or supply chain finance (Chung and Cárdenas-Barrón 2013; Guchhait, Maiti, and Maiti 2014), on the performance of the decision process. The tangible advantage of delayed payments is that suppliers give capital access and empower the downstream retailer to expand order sizes by preventing their business patterns to face a liquidity bottleneck. Equally, in this way, they not only help to increase the competitive position but also establish a lantern relationship with their business partners. Goyal (1985) demonstrated that the order quantity increases if predefined payment delays are allowed when contrasted with the traditional economic order quantity (EOQ) model. The replenishment decision under stock-dependent demand allowing a permissible delay in payments has been a notable topic of research in recent years (Teng et al. 2011).

Remark: Most of the studies exploring the influence of trade credit under stock-dependent demand primarily focused on the retailer's perspective, but it is more important to explore the decision under a multi-echelon supply chain setting to obtain realistic decisions.

Single vs. multiple items: The demand for an item in the store is also dependent on the inventory level for the other items on the shelf (Corstjens and Doyle 1981; Martínez-de-Albéniz and Roels 2011; Moon, Park, Hao, and Kim 2017). Therefore, it is reasonable to assume that the demand for each product will be affected negatively by the presence of others, that is, substitutable products and the presentation makes a positive effect (i.e., complementary products). Many firms, like departmental/supermarket managers, keep and order a group of items simultaneously, rather than individually. This class of problem is called "product assortment," a retailer that optimizes its shelf space allocation among a given selection of products and sometimes their prices. Urban (1998) introduced this class of problem and used a trial-and-error search technique to obtain a near-optimal decision. Most recently, Goyal, Levi, and Segev (2016) considered a single-period joint assortment and inventory-planning problem under dynamic substitution with stochastic demands.

Remark: Most of the studies explore replenishment decisions for single-item products under the display stock–dependent demand. Therefore, more studies are warranted by focusing on the effect of display stock for complementary vs. substitute products, and their effect on pricing and ordering decisions.

Note that the influence of time (Prasad and Mukherjee 2016), promotion or advertisement (Sana and Chaudhuri 2008), lead time (Michna, Nielsen, and Nielsen 2013), rebate (Dey et al. 2019), capacity constraint (Dye and Hsieh 2011), etc., are also important factors affecting replacement decisions under stock-dependent demand.

3.3 TYPES OF DEMAND FUNCTIONS AND THEIR ADVANTAGES AND DISADVANTAGES

In the last couple of decades, numerous theoretical display stock–dependent demand models have been developed to investigate the impact of display stock on consumer demand and study the firms' optimal replenishment decisions for single or multiple echelon supply chain settings (Huang, Leng, and Parlar 2013). It is possible to classify inventory-level-dependent demand into three main categories based on the limits for display quantity, as shown in Figure 3.2.

Moreover, it is possible to classify each component of demand into two categories: (1) additive, that is, linear function of display stock, and (2) multiplicative, that is, non-linear function of display stock along with other factors affecting demand function. Note that the impact of inventory depletion is not always steady with respect to time, in the later phase of the replenishment cycle; inventory depletion will have a higher diminishing effect. Therefore, the non-linear function may be closer to reality. Moreover, the stock-dependent demand is also categorized as sensitive to the initial inventory level and the instantaneous inventory level.

Note that coefficients a_i (>0), $i=1, 2, 3$ and b_i, (>0) $i=1, 2$ may be the constant demand rate independent of stock level or function of price (Saha and Goyal 2015), time (Saha, Das, and Basu 2012) and other variables. $\beta(>1)$ is the stock-sensitive demand parameter when demand is nonlinear, and $I(t)$ represents the level of display inventory at any time t. s_0 and s_1 represent upper and lower limits of display stock level.

In single-component demand (Chakraborty, Jana, and Roy 2015), it is assumed that the demand increases (or decreases) according to display stock level. Therefore, more inventory means more demand. This restricts reality in the following sense. First, from the perspective of increasing consumer demand, too much display of similar products can always create a negative impression of the product. Second, it is not feasible for a retailer to place all the replenishment quantity on the shelf, especially if items are deteriorating in nature. Finally, toward the end of the replenishment cycle, due to the lower amount of available quantity, display inventory becomes unable to make a similar impact on consumers compared to the beginning.

Consequently, the researchers propose a two-component demand (Yang 2014). The main motivation is to make the demand function more pragmatic to replicate the lower display effect during the end of the replenishment cycle. Therefore, it is assumed that if the level of display inventory reaches a certain level, s_1, then its impact on consumers remains constant. However, the major limitation both for single, that is, simply instantaneous inventory-level-dependent demand and two-component instantaneous inventory-level-dependent demand is that more stock on the shelf leads to more profit/lower cost. Therefore, in practice, it is hard to obtain a feasible optimal order quantity for a retailer under instantaneous inventory-level-dependent demand if holding cost or ordering cost is sufficiently low (Dye and Ouyang 2005). Chang, Goyal, and Teng (2006) pointed out this issue and argued that "too much piled up in everyone's way leaves a negative impression on buyer." The authors recommended an introduction of a constraint as an upper limit on the shelf because most retail outlets have limited shelf space. To overcome this limitation of inventory-level-dependent

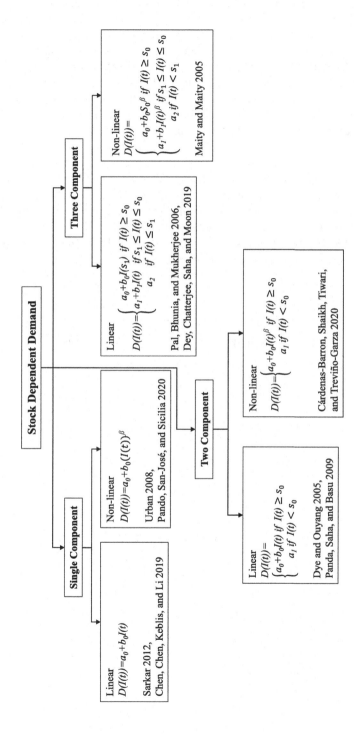

FIGURE 3.2 Variation of stock-dependent demand based on the display stock level.

demand, researchers proposed a market-oriented three-component demand function (Panda, Saha, and Goyal 2013). This group of authors argued that "in practice, the demand rate would not be dependent on display inventory level for the large stock. It would be displayed inventory level dependent within a range and beyond the range it is constant." Overall, this representation of three-component demand can be considered as a two-component stock-dependent demand with an upper limit measured on available shelf space. Consequently, it can be argued that three-component demand is the generalized version of the stock-dependent demand functions (Bhunia et al. 2014). As presented in Figure 3.2, the demand rate in a three-component demand function varies with the display stock level within the range s_0 to s_1. Outside this range, the demand rate becomes constant. If $s_0 \to \infty$ and $s_1 \to 0$, the three-component demand function is converted to single-component stack-dependent demand, which is discussed extensively in the literature (Sarkar 2012; Chen et al. 2019). If $s_1 \to \infty$, the demand function is converted to two-component stack-dependent demand, which is also common in the literature (Dye and Hsieh 2011).

Some of the functional forms used to characterize the display effect on the demand function with other relevant factors are presented in Table 3.1.

TABLE 3.1
Overview of Various Stock-Dependent Demands

Author	Demand type	Functional form
Sajadieh, Thorstenson, and Jokar 2010; Lee, Wang, and Chen 2017	Single-component non-linear in stock dependent demand	$D\big(I(t)\big) = aI(t)^{\beta}$ $a > 0, 0 < \beta < 1$
Chung and Cárdenas-Barrón 2013	Single-component linear in stock dependent demand	$D(t) = a + bI(t), \ a > 0, b > 0$
Saha, Das, and Basu 2012	Single-component linear stock, price, and exponential time-dependent demand	$D(t) = Ae^{-bt} + bI(t) - cp$
Chen et al. 2019	Single-component linear stock and price-dependent demand with demand decrease rate (μ)	$D_i(p,t) = \big[a(p) + bI_i(t)\big]^i, \ i = 1,2,\ldots,n$ $0 < \mu < 1$
Qin, Wang, and Wei 2014	Single-component linear stock, and generalize price ($H(p)$) and time ($N(t)$) dependent demand	$D\big(Q(t), N(t), p\big) = g\big(N(t)\big)H(p) - \theta I(t)$
Yang 2014,	Two-component non-linear stock-dependent demand	$D\big(I(t)\big) = \begin{cases} aI(t)^{\beta} & for \ I(t) > 0 \\ a & for \ I(t) \le 0 \end{cases}$ $a > 0, 0 \le \beta < 1$

Author	Demand type	Functional form
Chang, Teng, and Goyal 2010	Two-component linear stock-dependent demand	$D(I(t)) = \begin{cases} a+bI(t) \, for \, I(t) > 0 \\ \quad a \, for \, I(t) \leq 0 \end{cases}$
Shaikh et al. 2019	Two-component linear stock and price-dependent demand	$D(I(t)) = \begin{cases} a-cp+cI(t) \, for \, I(t) \geq 0 \\ \quad a-cp \, for \, I(t) < 0 \end{cases}$
Dye and Hsieh 2011	Linear in stock but non-linear in price	$D(I(t)) = \begin{cases} a(p)+bI(t) \, for \, I(t) > 0 \\ \quad a(p) \, for \, I(t) \leq 0 \end{cases}$
Chen et al. 2016	Two-component non-linear in stock and linear time-dependent demand	$D(I(t)) = \begin{cases} \alpha s_0{}^\beta \dfrac{m-t}{m} \, for \, 0 \leq t \leq t_1 \\ \alpha I(t)^\beta \dfrac{m-t}{m} \, for \, t_1 \leq t \leq T \end{cases}$
Duan et al. 2012	Two-component linear stock and partial backlogging dependent demand	$D(I(t)) = \begin{cases} a+bI(t) \, for \, I(t) \geq 0 \\ \dfrac{a+\delta_2 I(t)}{1+\delta_1(T-t)} \, for \, I(t) < 0 \end{cases}$
Feng, Chan, and Cárdenas-Barrón 2017	Two-component non-linear in stock, linear in time, and exponential price-dependent demand	$D(I(t)) = \begin{cases} \alpha s_0{}^\beta \dfrac{m-t}{m} e^{-\lambda p} \, for \, 0 \leq t \leq t_1 \\ \alpha I(t)^\beta \dfrac{m-t}{m} e^{-\lambda p} \, for \, t_1 \leq t \leq T \end{cases}$
Gupta, Bhunia, and Goyal 2007	Three-component linear in stock, price, and advertise (A) sensitive demand	$D(A,p,I) = \begin{cases} A^m(a-bp+cs_0) \, for \, I(t) > s_0 \\ A^m(a-bp+cI(t)) \, for \, s_1 < I(t) \leq s_0 \\ A^m(a-bp+cs_1) \, for \, 0 \leq I(t) \leq s_1 \\ \delta A^m(a-bp+cs_1) \, for \, I(t) < 0 \end{cases}$

Therefore, if the replenishment decision can be proposed under a three-component demand function, it can also be achievable to derive decisions for others. However, the exact values of s_0 and s_1 may vary from product to product. The single-component demand functions are widely used in the literature because it may lead to explicit results, and it is somewhat easy to estimate its parameters in an empirical study. However, for a supply chain setting, even a small change in demand information results in a significant change in profitability, thus more work is warranted in this direction.

3.4 AN OVERVIEW OF THE MODEL UNDER STOCK-DEPENDENT DEMAND

Although the literature on replenishment decisions under stock-dependent demand is widespread, most of the studies focus on the replenishment decision for a single firm or restrict their focus to only a single-component demand function if models are formulated under a supply chain setting. Along with the recognition of the importance of display stock level, researchers also initiated the pricing and replenishment decision in the supply chain framework. However, the work in this direction is limited. For example, Wang and Gerchak (2001) proposed models for coordinating decentralized two-stage supply chains when demand is shelf-space dependent. The authors explored the Nash equilibrium. Zhou, Min, and Goyal (2008) explored the coordination issue where the manufacturer adopts a lot-for-lot policy where demand is stock dependent. Yang and Zhang (2014) analyzed the effect of trade credit and quantity discounts from the perspective of achieving supply chain coordination under a stock-dependent demand rate. Saha and Goyal (2015) studied the effect of trade and direct rebate and wholesale price discount contracts. Chakraborty, Jana, and Roy (2015) formulated the model under a three-echelon supply chain, but they optimized centralized supply chain profit, not individual profits. The joint economic lot-sizing (JELS) problem is more popular under stock-dependent demand, and game-theoretic decision models are sanity. The idea of optimizing the joint total cost or profit in a single-vendor two-echelon supply chain model was considered early on by Goyal (1977), where the authors introduced the concept of JELS. In this direction, the work of Sajadieh, Thorstenson, and Jokar (2010); Giri and Bardhan (2015); and Michna, Disney, and Nielsen (2020) are worth mentioning. If looking to analyze the supply chain decision under a two- or three-component demand function, it is almost impossible to obtain a closed-form solution. Therefore, the optimal decision under the game-theoretic approach is still missing in the literature. To provide an overview in this regard, we formulate a simple model to determine the pricing-replenishment decision.

We consider a three-component linear demand function of the following form:

$$D\big(I(t),p\big) = \begin{cases} a + bs_0 - p, & 0 \le t \le t_1 \\ a + bI(t) - p, & t_1 \le t \le t_2 \\ a - p, & t_3 \le t \le T \end{cases} \tag{3.1}$$

where $I(t_1)=s_0$, $I(t_2)=s_1$, and $I(T)=0$. T represents cycle time. Consequently, we obtain the following differential equation governing the variation of inventory level at any time $t \in (0, T)$ as follows:

$$\frac{dI(t)}{dt} = -D\big(I(t),p\big), \ 0 \le t \le T \tag{3.2}$$

Therefore, after solving Equation (3.2), we can obtain the inventory level ($I(t)$) as follows:

$$I(t) = \begin{cases} Q - (a + bs_0 - p)t, & 0 \le t \le t_1 \\ s_0 e^{b(t_1-t)} + (a-p)\left(\dfrac{e^{b(t_1-t)}-1}{b}\right), & t_1 \le t \le t_2 \\ s_1 - (a-p)(t-t_2), & t_3 \le t \le T \end{cases} \tag{3.3}$$

where $t_1 = \dfrac{Q - s_0}{b s_0 + a - p}$, $t_2 = t_2 + \dfrac{1}{b} log \left[\dfrac{b s_0 + a - p}{b s_1 + a - p} \right]$, and $T = t_2 + \dfrac{s_1}{a - p}$

Therefore, if h represents per-unit holding cost for the retailer side, the total holding cost (HC) can be obtained as follows:

$$HC = h \left[\int_0^{t_1} I(t) dt + \int_{t_1}^{t_2} I(t) dt + \int_{t_3}^{T} I(t) dt \right]$$

$$= h \left[\dfrac{Q^2 - s_0^2}{2(b s_0 + a - p)} + \dfrac{s_1^2}{2(a - p)} \right.$$

$$\left. + \left(\dfrac{s_0 - s_1}{b} - \dfrac{(a - p)}{b^2} log \left[\dfrac{b s_0 + a - p}{b s_1 + a - p} \right] \right) \right]$$

Consequently, the total profit per unit time (TPU) for the retailer and the manufacturer can be presented as follows:

$$Maximize\, TPU_r (Q, p) = \dfrac{(p - w)Q - HC - A_r}{T} \tag{3.4}$$

$$Maximize\, TPU_m (Q, p) = \dfrac{wQ - A_m}{T} \tag{3.5}$$

where A_r, A_m, and w represent ordering/setup costs for the retailer and manufacturer and wholesale price for the manufacturer. It can be noted that the objective function is non-linear in nature due to the presence of the logarithm function. Besides this, the total cycle time is also a function of price and initial order quantity as follows:

$$T(Q, p) = \dfrac{Q - b s_0}{b s_0 + a - p} + \dfrac{1}{b} log \left[\dfrac{b s_0 + a - p}{b s_1 + a - p} \right] + \dfrac{s_1}{a - p} \tag{3.6}$$

Therefore, the optimal decision for the retailer will be obtained by solving the following two equations simultaneously.

$$\dfrac{\partial TPU}{\partial p} = 0 \Rightarrow \dfrac{TP}{T} = \dfrac{\partial TP}{\partial p} \bigg/ \dfrac{\partial T}{\partial p} \tag{3.7}$$

$$\dfrac{\partial TPU}{\partial Q} = 0 \Rightarrow \dfrac{TP}{T} = \dfrac{\partial TP}{\partial Q} \bigg/ \dfrac{\partial T}{\partial Q} = ((p - w)(b s_0 + a - \beta p) - hQ) \tag{3.8}$$

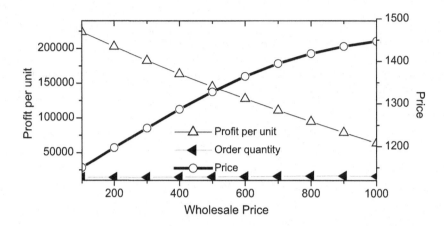

FIGURE 3.3 Profit per unit time, order quantity, and product price.

From the expression on the right-hand side in Equation (3.8), the importance of the work by Chang, Goyal, and Teng (2006) can be noted in this direction; the upper limit on shelf space can ensure the optimality. However, the main point is that by solving differential Equations (3.7) and (3.8), it is impossible to find the closed-form representation without approximation. Consequently, as mentioned earlier, the optimal decision under a generalized stock-dependent demand, that is, under a two- or three-component demand function is still missing in the existing literature under the game-theoretic framework. After solving the problem from the perspective of a retailer, we present optimal decisions in the following figure:

From Figure 3.3, it is obvious that the retailer needs to adjust the retail price progressively with the increment of the wholesale price. The total profit-per-unit time decreases, but the retailer needs to order more to compensate for the cost. Therefore, under a non-cooperative supply chain setting or the influence of the coordination contract mechanism, it is important to explore the nature of the optimal decision.

It should be noted that we ignore factors such as product deterioration, partial backlogging or non-zero end inventory level or backlogging, trade credit, dynamic pricing, and reference price effect, which are relevant. It is difficult to obtain the closed-form decision, which is important for exploring the nature of optimal decisions under the game-theoretic framework. Therefore, it is necessary to apply some sort of algorithm expertise to solve such a complex decision-making model.

3.5 EXAMPLES OF OPTIMIZATION MODELS

In accordance with the discussion in the previous section, we can realize that finding optimal decisions under the influence of stock-dependent demand is not straightforward. To solve the problem from the perspective of a single firm, researchers primarily rely on the iterative algorithm (Chung and Cárdenas-Barrón 2013; Yang 2014; Shaikh et al. 2019), sometimes mixed with the numerical methods, such as the Newton–Raphson and Runge–Kutta methods. Moreover, researchers have used commercial software such as Mathematica (Maihami and Abadi 2012; Giri and Bardhan

2015), Maple (Mishra, Singh, and Kumar 2013), and Lingo (Lee, Wang, and Chen 2017). In this direction, some authors use a problem-specific heuristic to reduce computational complexity (Olsson 2019). Recently, researchers employed some standard metaheuristics, such as the particle swarm optimization (PSO) technique employed by Bhunia et al. (2014), while determining optimal replenishment quantity under the joint effect of product deterioration and partial backlogging. Tiwari et al. (2017) also used PSO to determine the optimal order quantity while integrating the effect of the non-instantaneous deteriorating product. Similarly, the genetic algorithm (GA) is employed by Guchhait, Maiti, and Mai (2014) and Mondal, Maiti, and Maiti (2014) to find the optimal order quantity for a retailer facing three-component stock-dependent demand for the deteriorating item, and Chakraborty, Jana, and Roy (2015) to find an optimal decision for a two-echelon supply chain facing linear stock-dependent demand. Önal, Yenipazarli, and Kundakcioglu (2016) used the tabu search (TS) technique while determining space-allocation decisions for a group of perishable products under stock-dependent demand. A hybrid bat algorithm (HBA) is proposed by Dey et al. (2019) to obtain a dynamic pricing strategy for a retailer as well as a rebate scheme under three-component stock-dependent demand. Maiti and Maiti (2005) used the simulated annealing (SA) algorithm with a retailer facing three-component non-linear demand. Overall, replenishment decisions under the stock-dependent demand for multiple items can lead to an NP-hard (nondeterministic polynomial time) problem. Therefore, simulation methods can be adopted to evaluate inventory-pricing policy instead of an optimal solution to obtain deep insight, especially in a supply chain setting.

3.6 INDUSTRIES AND PRODUCTS AFFECTED BY THE FACTORS

As stated by Whitin (1957), "For retail stores, the inventory control problem for style goods is further complicated by the fact that inventory and sales are not independent of one another." There are plenty of items that experience inventory-dependent demand, which is what the consumers see on the shelf; therefore, sensitivity to stock is more applicable for business-to-consumer (B2C). For example, low inventory levels for apparel or packaged goods can decrease demand significantly, and it is referred to as the "broken assortment" effect. In recent years, modern retailers have increasingly benefited from the use of inventory-tracking technologies to improve operational efficiency and take advantage of the billboard and scarcity effects of inventories. There are examples of such conditions in electronics, automotive, and designer apparel industries where firms deliberately restrict the supply of their products to drive up demand (Xue et al. 2017). It is commonly presumed that the display item is the retailer's primary point of contact with customers and a retailer keeps the space for the most valuable resources. On the operational side, there is naturally a cost difference between carrying an item in the backroom vs. on the retail shelf. These differences have only recently attracted attention from the inventory management literature (Yang and Zhang 2014). If the demand is relatively stable, then the retailer can use a fixed shelf inventory policy and keep additional products in the store to reduce holding costs as well as deterioration rate. If demand were non-stationary in nature, then dynamic adjustment of the on-shelf inventory levels and pricing decision would be more practical. On the other hand, display stock and product deterioration are not independent. According to Buzby, Farah-Wells, and Hyman (2014), the expected total value of food product loss at the retail end in 2010 in the United States was $53.8 billion, primarily due to continuous deterioration. Therefore, joint dynamic pricing and the proper display are important tools for the retailer, not only to reduce such a financial loss but also to achieve future sustainability goals.

3.7 CONCLUSION AND FUTURE RESEARCH DIRECTIONS

In conclusion, we find that the display stock–dependent demand model proves to be useful in analyzing the replenishment decision model and has the potential to generate possibilities for developing new types of demand functions and solution methodology that can further be useful for other problems. The following are the key directions to be explored:

First, in academic literature and available retail software tools, some important solutions exist to minimize difficulties for day-to-day operations. However, there are limited options for the store managers to determine when and what products are to be selected for placement on their promotional display space, and disposal decisions, mainly for perishable inventory systems with the priority classes from the perspective of consumers. Therefore, more work is required in this direction. Second, in a competitive market, prices of products are always a key factor that affects consumer decisions. Consequently, besides determining shelf space limits, replenishment quantity, and price separately for multiple products, a decision support system is always not only important for the retail managers; from the perspective of the overall supply chain operations, flexibility and scalability to incorporate day-to-day operations is always a topic of research interest. Third, assortment planning receives considerable attention from researchers to propose a policy to achieve commercial success for a firm and its associated supply chain members. In these circumstances, the retailer's binary decision to display products can affect the whole supply chain. Although the problem leads to complex NP-hard decision-making scenarios under the influence of the cross-selling and cross-price effect, dynamic product substitution, shelf-space capacity constraints, and consumer loyalty with a brand not only affect the financial benefits but also the long-term success. In this regard, exploring heuristic approaches to obtain near-optimal quantization and pricing decisions can be a new research trend. Finally, and most importantly, the optimal analytical decision under game-theoretic approaches or a supply chain coordination contract is missing under a two- or three-component stock-dependent demand function that incorporates the effect of reference price, multiple items, product deterioration, dynamic pricing scheme, backlogging, etc. Therefore, more studies are required in this direction.

REFERENCES

Alfares, H. K. 2007. Inventory model with stock-level dependent demand rate and variable holding cost. *International Journal of Production Economics* 108, no. 1–2: 259–265.

Bakker, M., J. Riezebos, and R. H. Teunter. 2012. Review of inventory systems with deterioration since 2001. *European Journal of Operational Research* 221, no. 2: 275–284.

Balakrishnan, A., M. S. Pangburn, and E. Stavrulaki. 2004. "Stack them high, let'em fly": Lot-sizing policies when inventories stimulate demand. *Management Science* 50, no. 5: 630–644.

Bhunia, A. K., S. K. Mahato, A. A. Shaikh, and C. K. Jaggi. 2014. A deteriorating inventory model with displayed stock-level-dependent demand and partially backlogged shortages with all unit discount facilities via particle swarm optimization. *International Journal of Systems Science: Operations & Logistics* 1, no. 3: 164–180.

Buzby, J. C., H. Farah-Wells, and J. Hyman. 2014. The estimated amount, value, and calories of postharvest food losses at the retail and consumer levels in the United States. *USDA-ERS Economic Information Bulletin* 121.

Cachon, G. P., S. Gallino, and M. Olivares. 2019. Does adding inventory increase sales? Evidence of a scarcity effect in US automobile dealerships. *Management Science* 65, no. 4: 1469–1485.

Cárdenas-Barrón, L. E., A. A. Shaikh, S. Tiwari, and G. Treviño-Garza. 2020. An EOQ inventory model with nonlinear stock dependent holding cost, nonlinear stock dependent demand and trade credit. *Computers & Industrial Engineering* 139: 105557.

Chakraborty, D., D. K. Jana, and T. K. Roy. 2015. Multi-item integrated supply chain model for deteriorating items with stock dependent demand under fuzzy random and bifuzzy environments. *Computers & Industrial Engineering* 88: 166–180.

Chang, C. T. 2004. Inventory models with stock-dependent demand and nonlinear holding costs for deteriorating items. *Asia-Pacific Journal of Operational Research* 21, no. 4: 435–446.

Chang, C. T., S. K. Goyal, and J. T. Teng. 2006. On "an EOQ model for perishable items under stock-dependent selling rate and time-dependent partial backlogging" by Dye and Ouyang. *European Journal of Operational Research* 174, no. 2: 923–929.

Chang, C. T., J. T. Teng, and S. K. Goyal. 2010. Optimal replenishment policies for non-instantaneous deteriorating items with stock-dependent demand. *International Journal of Production Economics* 123, no. 1: 62–68.

Chen, L., X. Chen, M. F. Keblis, and G. Li. 2019. Optimal pricing and replenishment policy for deteriorating inventory under stock-level-dependent, time-varying and price-dependent demand. *Computers & Industrial Engineering* 135: 1294–1299.

Chen, S. C., J. Min, J. T. Teng, and F. Li. 2016. Inventory and shelf-space optimization for fresh produce with expiration date under freshness-and-stock-dependent demand rate. *Journal of the Operational Research Society* 67, no. 6: 884–896.

Chung, K. J., and L. E. Cárdenas-Barrón. 2013. The simplified solution procedure for deteriorating items under stock-dependent demand and two-level trade credit in the supply chain management. *Applied Mathematical Modelling* 37, no. 7: 4653–4660.

Corstjens, M., and P. Doyle. 1981. A model for optimizing retail space allocations. *Management Science* 27, no. 7: 822–833.

Dehghanbaghi, N., and M. S. Sajadieh. 2017. Joint optimization of production, transportation and pricing policies of complementary products in a supply chain. *Computers & Industrial Engineering* 107: 150–157.

Desmet, P., and V. Renaudin. 1998. Estimation of product category sales responsiveness to allocated shelf space. *International Journal of Research in Marketing* 15, no. 5: 443–457.

Dey, K., D. Chatterjee, S. Saha, and I. Moon. 2019. Dynamic versus static rebates: An investigation on price, displayed stock level, and rebate-induced demand using a hybrid bat algorithm. *Annals of Operations Research* 279, no. 1–2: 187–219.

Duan, Y., G. Li, J. M. Tien, and J. Huo. 2012. Inventory models for perishable items with inventory level dependent demand rate. *Applied Mathematical Modelling* 36, no. 10: 5015–5028.

Dye, C. Y., and T. P. Hsieh. 2011. Deterministic ordering policy with price-and stock-dependent demand under fluctuating cost and limited capacity. *Expert Systems with Applications* 38, no. 12: 14976–14983.

Dye, C. Y., and L. Y. Ouyang. 2005. An EOQ model for perishable items under stock-dependent selling rate and time-dependent partial backlogging. *European Journal of Operational Research* 163, no. 3: 776–783.

Feng, L., Y. L. Chan, and L. E. Cárdenas-Barrón. 2017. Pricing and lot-sizing polices for perishable goods when the demand depends on selling price, displayed stocks, and expiration date. *International Journal of Production Economics* 185: 11–20.

Ghare, P. M., and G. F. Schrader. 1963. A model for an exponential decaying inventory. *Journal of Industrial Engineering* 14: 238–243.

Giri, B. C., and S. Bardhan. 2015. A vendor—buyer JELS model with stock-dependent demand and consigned inventory under buyer's space constraint. *Operational Research* 15, no. 1: 79–93.

Goyal, S. K. 1977. An integrated inventory model for a single supplier-single customer problem. *International Journal of Production Research* 15, no. 1: 107–111.

Goyal, S. K. 1985. Economic order quantity under conditions of permissible delay in payments. *Journal of the Operational Research Society* 36, no. 4: 335–338.

Goyal, S. K., and C. T. Chang. 2009. Optimal ordering and transfer policy for an inventory with stock dependent demand. *European Journal of Operational Research* 196, no. 1: 177–185.

Goyal, V., R. Levi, and D. Segev. 2016. Near-optimal algorithms for the assortment-planning problem under dynamic substitution and stochastic demand. *Operations Research* 64, no. 1: 219–235.

Guchhait, P., M. K. Maiti, and M. Maiti. 2014. Inventory policy of a deteriorating item with variable demand under trade credit period. *Computers & Industrial Engineering* 76: 75–88.

Gupta, R. K., A. K. Bhunia, and S. K. Goyal. 2007. An application of genetic algorithm in a marketing oriented inventory model with interval valued inventory costs and three-component demand rate dependent on displayed stock level. *Applied Mathematics and Computation* 192, no. 2: 466–478.

Hsu, P. H., H. M. Wee, and H. M. Teng. 2010. Preservation technology investment for deteriorating inventory. *International Journal of Production Economics* 124, no. 2: 388–394.

Huang, J., M. Leng, and M. Parlar. 2013. Demand functions in decision modeling: A comprehensive survey and research directions. *Decision Sciences* 44, no. 3: 557–609.

Khan, M. A. A., A. A. Shaikh, G. C. Panda, I. Konstantaras, and L. E. Cárdenas-Barrón. 2020. The effect of advance payment with discount facility on supply decisions of deteriorating products whose demand is both price and stock dependent. *International Transactions in Operational Research* 27, no. 3: 1343–1367.

Koschat, M. A. 2008. Store inventory can affect demand: Empirical evidence from magazine retailing. *Journal of Retailing* 84, no. 2: 165–179.

Krommyda, I. P., K. Skouri, and I. Konstantaras. 2015. Optimal ordering quantities for substitutable products with stock-dependent demand. *Applied Mathematical Modelling* 39, no. 1: 147–164.

Lee, W., S. P. Wang, and W. C. Chen. 2017. Forward and backward stocking policies for a two-level supply chain with consignment stock agreement and stock-dependent demand. *European Journal of Operational Research* 256, no. 3: 830–840.

Levin, R. I. 1972. *Production Operations Management: Contemporary Policy for Managing Operating Systems*. New York: McGraw-Hill Companies.

Maihami, R., and I. N. K. Abadi. 2012. Joint control of inventory and its pricing for non-instantaneously deteriorating items under permissible delay in payments and partial backlogging. *Mathematical and Computer Modelling* 55, no. 5–6: 1722–1733.

Maiti, M. K., and M. Maiti. 2005. Inventory of damageable items with variable replenishment and unit production cost via simulated annealing method. *Computers & Industrial Engineering* 49, no. 3: 432–448.

Martínez-de-Albéniz, V., and G. Roels. 2011. Competing for shelf space. *Production and Operations Management* 20, no. 1: 32–46.

Michna, Z., S. M. Disney, and P. Nielsen. 2020. The impact of stochastic lead times on the bullwhip effect under correlated demand and moving average forecasts. *Omega* 93: 102033.

Michna, Z., I. E. Nielsen, and P. Nielsen. 2013. The bullwhip effect in supply chains with stochastic lead times. *Mathematical Economics* 9, no. 16: 71–88.

Min, J., Y. W. Zhou, and J. Zhao. 2010. An inventory model for deteriorating items under stock-dependent demand and two-level trade credit. *Applied Mathematical Modelling* 34, no. 11: 3273–3285.

Mishra, V. K., L. S. Singh, and R. Kumar. 2013. An inventory model for deteriorating items with time-dependent demand and time-varying holding cost under partial backlogging. *Journal of Industrial Engineering International* 9, no. 1: 4.

Mondal, M., M. K. Maiti, and M. Maiti. 2014. A two storage production-repairing model with fuzzy defective rate and displayed inventory dependent demand. *Optimization and Engineering* 15, no. 3: 751–772.

Moon, I., K. S. Park, J. Hao, and D. Kim. 2017. Joint decisions on product line selection, purchasing, and pricing. *European Journal of Operational Research* 262, no. 1: 207–216.

Olsson, F. 2019. Simple modeling techniques for base-stock inventory systems with state-dependent demand rates. *Mathematical Methods of Operations Research* 90, no. 1: 61–76.

Önal, M., A. Yenipazarli, and O. E. Kundakcioglu. 2016. A mathematical model for perishable products with price-and displayed-stock-dependent demand. *Computers & Industrial Engineering* 102: 246–258.

Padmanabhan, G., and P. Vrat. 1995. EOQ models for perishable items under stock dependent selling rate. *European Journal of Operational Research* 86, no. 2: 281–292.

Pal, A. K., A. K. Bhunia, and R. N. Mukherjee. 2006. Optimal lot size model for deteriorating items with demand rate dependent on displayed stock level (DSL) and partial backordering. *European Journal of Operational Research* 175, no. 2: 977–991.

Panda, S., S. Saha, and M. Basu. 2009. An EOQ model for perishable products with discounted selling price and stock dependent demand. *Central European Journal of Operations Research* 17, no. 1: 31.

Panda, S., S. Saha, and S. K. Goyal. 2013. Dilemma of rented warehouse and shelf for inventory systems with displayed stock level dependent demand. *Economic Modelling* 32: 452–462.

Pando, V., L. A. San-José, and J. Sicilia. 2019. Profitability ratio maximization in an inventory model with stock-dependent demand rate and non-linear holding cost. *Applied Mathematical Modelling* 66: 643–661.

Pando, V., L. A. San-José, and J. Sicilia. 2020. A new approach to maximize the profit/cost ratio in a stock-dependent demand inventory model. *Computers & Operations Research*: 104940.

Prasad, K., and B. Mukherjee. 2016. Optimal inventory model under stock and time dependent demand for time varying deterioration rate with shortages. *Annals of Operations Research* 243, no. 1–2: 323–334.

Qin, Y., J. Wang, and C. Wei. 2014. Joint pricing and inventory control for fresh produce and foods with quality and physical quantity deteriorating simultaneously. *International Journal of Production Economics* 152: 42–48.

Ritchie, E., and A. Tsado. 1985. Stock replenishment quantities for unbounded linear increasing demand: An interesting consequence of the optimal policy. *Journal of the Operational Research Society* 36, no. 8: 737–739.

Saha, S., S. Das, and M. Basu. 2012. Supply chain coordination under stock-and price-dependent selling rates under declining market. *Advances in Operations Research* 2012: 0–14.

Saha, S., and S. K. Goyal. 2015. Supply chain coordination contracts with inventory level and retail price dependent demand. *International Journal of Production Economics* 161: 140–152.

Sajadieh, M. S., A. Thorstenson, and M. R. A. Jokar. 2010. An integrated vendor—buyer model with stock-dependent demand. *Transportation Research Part E: Logistics and Transportation Review* 46, no. 6: 963–974.

Sana, S. S., and K. S. Chaudhuri. 2008. An inventory model for stock with advertising sensitive demand. *IMA Journal of Management Mathematics* 19, no. 1: 51–62.

San-José, L. A., J. Sicilia, M. González-De-la-Rosa, and J. Febles-Acosta. 2019. Analysis of an inventory system with discrete scheduling period, time-dependent demand and backlogged shortages. *Computers & Operations Research* 109: 200–208.

Sarkar, B. 2012. An EOQ model with delay in payments and stock dependent demand in the presence of imperfect production. *Applied Mathematics and Computation* 218, no. 17: 8295–8308.

Shaikh, A. A., M. A. A. Khan, G. C. Panda, and I. Konstantaras. 2019. Price discount facility in an EOQ model for deteriorating items with stock-dependent demand and partial backlogging. *International Transactions in Operational Research* 26, no. 4: 1365–1395.

Silver, E. A., and H. C. Meal. 1969. A simple modification of the EOQ for the case of a varying demand rate. *Production and Inventory Management* 10, no. 4: 52–65.

Teng, J. T., I. P. Krommyda, K. Skouri, and K. R. Lou. 2011. A comprehensive extension of optimal ordering policy for stock-dependent demand under progressive payment scheme. *European Journal of Operational Research* 215, no. 1: 97–104.

Tiwari, S., C. K. Jaggi, A. K., Bhunia, A. A. Shaikh, and M. Goh. 2017. Two-warehouse inventory model for non-instantaneous deteriorating items with stock-dependent demand and inflation using particle swarm optimization. *Annals of Operations Research* 254, no. 1–2: 401–423.

Urban, T. L. 1998. An inventory-theoretic approach to product assortment and shelf-space allocation. *Journal of Retailing* 74, no. 1: 15–35.

Urban, T. L. 2005. Inventory models with inventory-level-dependent demand: A comprehensive review and unifying theory. *European Journal of Operational Research* 162, no. 3: 792–804.

Urban, T. L. 2008. An extension of inventory models with discretely variable holding costs. *International Journal of Production Economics* 114, no. 1: 399–403.

Wang, Y., and Y. Gerchak. 2001. Supply chain coordination when demand is shelf-space dependent. *Manufacturing & Service Operations Management* 3, no. 1: 82–87.

Whitin, T. M. 1957. *The Theory of Inventory Management*. Princeton, NJ: Princeton University Press.

Xue, W., O. Caliskan Demirag, F. Y. Chen, and Y. Yang. 2017. Managing retail shelf and backroom inventories when demand depends on the shelf-stock level. *Production and Operations Management* 26, no. 9: 1685–1704.

Yang, C. T. 2014. An inventory model with both stock-dependent demand rate and stock-dependent holding cost rate. *International Journal of Production Economics* 155: 214–221.

Yang, H. L., J. T. Teng, and M. S. Chern. 2010. An inventory model under inflation for deteriorating items with stock-dependent consumption rate and partial backlogging shortages. *International Journal of Production Economics* 123, no. 1: 8–19.

Yang, N., and R. Zhang. 2014. Dynamic pricing and inventory management under inventory-dependent demand. *Operations Research* 62, no. 5: 1077–1094.

Zhang, J., Y. Wang, L. Lu, and W. Tang. 2015. Optimal dynamic pricing and replenishment cycle for non-instantaneous deterioration items with inventory-level-dependent demand. *International Journal of Production Economics* 170: 136–145.

Zhou, Y. W., J. Min, and S. K. Goyal. 2008. Supply-chain coordination under an inventory-level-dependent demand rate. *International Journal of Production Economics* 113, no. 2: 518–527.

4 Rebate Contracts

F. Sadeghi
Department of Industrial Engineering, Jam Faculty of
Engineering, Persian Gulf University, Bushehr, Iran

M. Hemmati
Department of Industrial Engineering, Amirkabir
University of Technology, Tehran, Iran

CONTENTS

4.1 INTRODUCTION

Rebates have become one of the most important and main advertising tools of manufacturers and retailers over the last decade (Li et al. 2016). Rebates have two main benefits. First, a rebate can keep customer expectations of future product prices while increasing demand (Khouja 2006, Khouja and Zhou 2010). By using a temporary discount per product, customers can help the store keep the price of the product constant, save on their purchase costs, and take advantage of a manufacturer or retailer rebate. Second, a rebate can reduce costs more than price markdowns due to low claim rates (Cho, McCardle, and Tang 2009). Consumers may not redeem cash for various reasons, such as not being sensitive to the rebate, forgetting to redeem, or long redemption time. As a result, the first offer of a rebate garners more profits (Li et al. 2016).

A rebate is used in different titles and modes like mail-in rebate (MIR), coupons, instant rebate, and retailer rebate, also known as a channel rebate. The most common rebate is MIR in which customers, generally by following specific rules, can redeem cash by gathering paperwork, filling out forms, and submitting rebate requests within a specific time frame (Li et al. 2016). Briefly, MIR incentivizes consumers to buy a product or collections of products by offering them cash (Geng and Mallik 2011). MIR serves as a practical tool in many industries that have an important relationship with short seasonal products, such as Nikon digital cameras (Muzaffar, Malik, and Rashid 2018). With a retailer rebate, the manufacturer, according to the quantity ordered, passes the rebate to the retailer, and then the retailer can pass part of it to the final customers (Muzaffar, Malik, and Rashid 2018).

There are different executions for both retailer rebates and MIR. Retailer rebates are paid for products of which the retailer sells more than the target level (Muzaffar, Malik, and Rashid 2018). Also, MIRs are prevalent in customer products like software and electronics or home appliances and cosmetics (Geng and Mallik 2011).

Rebate value is the essential factor in wholesale transactions and has an effective role in gaining profit or loss in business. Many inventory models deal with different types of demand functions and rebates (Valliathal and Uthayakumar 2011).

Rebates are studied in both marketing and operations, broadly. Marketing considers price and consumer choice under a rebate, while operations focuses on the overall supply chain (SC) dynamics involving inventory and the profit implication of rebates. Considering the operations management point of view, two main types of rebate, the sales rebate and consumer rebate, are studied. The sales rebate changes from a manufacturer to a retailer when the specific circumstances for sales are provided. The consumer rebate, on the other hand, goes straight to the customer (Geng and Mallik 2011).

It should also be considered that the area of operations management is prosperous in models that investigate the amount of common orders and pricing decisions. In these types of models, management manipulates demand by changing product unit cost. Also, management can raise demand by incremental marketing costs. As the demand rate has a direct influence on batch sizing, models were developed to specify the optimal price, batch size, and marketing costs cooperatively. MIR is another way in which manufacturers and retailers can affect demand. Hence, management has an additional decision variable to bring under control: offering the optimal rebate value (Khouja 2006).

This chapter aims to study the effect of rebate contracts on demand from an operations management point of view. Therefore, the explanation and citation of other types of rebates are avoided.

First, the conceptual framework of rebate contracts is described, then the rebate-sensitive demand functions, their cons, and their pros are presented. Some related case studies and examples of optimization models are then reviewed. Finally, the research trends and suggestions for future studies are presented, respectively.

4.2 CONCEPTUAL FRAMEWORK

It has been well observed that manufacturers and retailers, jointly or individually, offer a huge amount of rebates on a routine posted price with the aim of increasing item demand (Saha and Sarmah 2013). The effect of different types of rebates on demand is very wide and is studied in many papers.

Today, a highly competitive environment prevails in the industries; managing different but affiliate members of the SC are the most important summons of SC systems. Coordination models and the stimulus plans are implemented via the SC

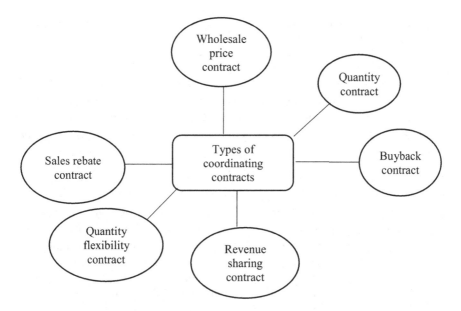

FIGURE 4.1 Types of coordination contracts.

contracts, which are collections of obligations and orders. These contracts offer a stimulus system (risk and reward sharing) also applied data to ensure that all members of the SC are in line with the goals of the whole SC (Heydari and Asl-Najafi 2021). Figure 4.1 shows the type of coordination contracts in SC.

Revenue sharing: In this contract, the supplier offers the goods to the retailer at a lower wholesale price, and in return, the retailer gives part of its revenue to the supplier (Heydari and Asl-Najafi 2021).

Quantity flexibility: The supplier permits the retailer to regulate their order quantity after knowing the exact amount of market demand. In this type of contract, the retailer orders more than the specified quantity; in return, the supplier provides a definite amount more than the initial amount of the retailer, if necessary.

Quantity discount: In some cases, a quantity discount contract is like a sales rebate. The supplier regards a discount for ordering more than the specified amount, although the sales rebate agreement deals with two major parameters, sales target level and rebate amount (Heydari and Asl-Najafi 2021).

Wholesale price: In an SC, manufacturers and retailers attempt to balance supply and demand by using production planning and demand management. In the case of seasonal products and the fashion apparel industry, manufacturers may try different methods and strategies to boost SC efficiency as well as reduce the pressure of product shortage. Wholesale price is one of the main methods used in this situation. To reduce the likelihood of shortages, manufacturers can start the production process sooner and encourage retailers to maintain these productions. To achieve this goal, retailers must be properly motivated to recoup the cost of sorting these productions. The manufacturers can use wholesale prices to suggest grants to the retailer for taking delivery in the earlier period (K.-L. Huang, Kuo, and Lu 2014); hence, in this method, the manufacturer increases the wholesale price applied to the retailer to a high level and the retailer receives more benefits when the order quantity is greater than the single-period capacity.

Trade credit: The purchase of commodities and services usually needs players to consent in trade terms; this mostly includes prices, delivery, and payment circumstances (Seifert, Seifert, and Protopappa-Sieke 2013). In capital-limited SCs (Zhan, Chen, and Hu 2019), while payment circumstances are likely to be very simple, it is possible for firms to allow payment delays. The optimal period of these payment delays is a major concern of credit term decisions (Seifert, Seifert, and Protopappa-Sieke 2013).

Trade credit, which is also known as permissible delay in payment (Cárdenas-Barrón et al. 2020), is applied as short-term financing by many companies to extend their business capacity and attract more consumers (Shaikh et al. 2019). Trade credit is a promotional method and a means of competition that can help players build a good, long-term relationship (Yang, Hong, and Lee 2014). Moreover, it also affects the customer's ordering strategy and allows them to invest additional liquidity elsewhere and obtain more profit (Heydari, Rastegar, and Glock 2017).

Full trade credit and partial trade credit are the two main groups of credit policies. In the former, the firm gives the credit on the buyer's total purchase, and in the latter, the firm gives the credit on a fraction of the buyer's total purchase. Moreover, conditional trade credit, or one-level credit, in which only the wholesaler suggests the credit to the retailer, and two-level credit, in which both retailer and wholesaler recommend credit to retailer and consumer, respectively, are other types of credit periods (Pramanik, Maiti, and Maiti 2017).

Effective use of this credit requires the wisdom to avoid unnecessary expenses and take advantage of the credit by reducing capital dependence on other resources (Giri, Bhattacharjee, and Maiti 2018). Uncertainties in companies' exchanges of commodity cause uncertainties in money flow, so stochastic money maintains expenditure. The company abates these expenditures by setting payment terms and providing information about money requirements and expected receipts.

There are two main classes of the economic order quantity (EOQ) that concern credit policy in the literature. In the first class, inventory models with no deterioration and no shortage are considered, while the second class includes inventory models with deterioration and no shortage (Seifert, Seifert, and Protopappa-Sieke 2013).

Buyback: It is obvious that because of more profits, the wholesale price–only contracts generally cause some disruptions and anomalies in SC performance in the face of uncertain demand and market demand. Many contracting systems have been extended in SC management to reduce efficiency degradation. A common type is the buyback mechanism. In this type of contract, the retailer still yields the wholesale price for each unit of orders; however, it is possible to give back all or some of the unsold products to the suppliers with a prearranged full or partial refund per unit at the end of the sales season. Buyback contracts are widely used in different retail divisions, including publishing, fashion and clothing, electronics, and beauty (Zhao et al. 2014).

Sales rebates: Through this contract, the supplier gives a certain amount of rebate to the retailer for each unit sold more than the sales target level. Because the sales rebate contract focuses on sales amount rather than order quantity, it is different from quantity discount agreements. Also, considering that the sales rebate contract has a straight influence on the retailers to sell more, it is more productive and effective than the quantity discount agreement (Yang et al. 2015).

In the sales rebate contract, based on which retailers sell more, the supplier will pay a certain amount of the rebate, which is the essential reason for the retailers to directly use procedures like sales efforts, promotions, and discounted retail prices to raise the entire sales amount. To create a straight stimulus for retailers to raise their sales amount, a sales rebate, in some cases, is a more productive agreement than other similar agreements. Hence, sales rebates can be convenient for SCs to obtain coordination by managing issues like double marginalization (Yang et al. 2015).

Liner and target sales rebates are two prevalent types of sales rebates. In the first one, the manufacturer pays a fixed rebate per unit sold to the retailer, regardless of the quantity sold (Chiu et al. 2012). The second one is paid for each unit sold more than the predetermined target sales rebate. So, the amount of the rebate is a function of the retailer's particular sales performance according to the sales' amount levels (Chiu et al. 2012).

All in all, sales rebate contracts are considered a powerful and effective incentive tool to make retailers sell more (Chiu et al. 2012).

Mail-in rebate: MIR is a prevalent advertising tool applied in marketing consumer products and/or services. Consumers are offered a delayed stimulus of cash (or gift cards) on the purchase of an item, a collection of items, or an upgrade of an item. MIRs are frequently used in a variety of products extending from software and electronics to home appliances and cosmetics. The MIRs process for redeeming cash is generally as follows: gathering paperwork, filling out the forms, cutting the Universal Product Code (UPC), and submitting the rebate application within a particular amount of time (Geng and Mallik 2011).

After the retailers or their collecting agents process the rebate forms, the refunds are sent to the consumers as a check or gift card. Moreover, both the consumers and the retailers are charged with non-insignificant costs in the redemption process (Hu, Hu, and Ye 2017).

Instant rebates: This rebate is applied at the time of purchase with low implementation cost for retailers, such as a yellow price tag and an immediate discount (Hu, Hu, and Ye 2017).

Consumer rebates: This type of contract, which is not less popular than retailer rebates, includes payments from the manufacturer to the consumer after the consumer buys the manufacturer's product.

Retailer rebates: This rebate is also known as the channel rebate in which payments from the manufacturer to the retailer are based on the sales performance of the retailer (Agrawal and Smith 2010). Retailer rebates can also be paid for each unit the retailer sells to the final customers or only for units that are sold after the retailer reaches the target amount (Agrawal and Smith 2010).

For both retailer and consumer rebates, there are several implementations. Retailer rebates go from the manufacturer to the consumer for each unit they sell. However, a consumer rebate goes from the manufacturer to the customers for each unit they purchase (Agrawal and Smith 2010).

4.3 DEMAND FUNCTION

Manufacturers and retailers try to manipulate the demand for their product by offering rebates to the final customers. Indeed, the manufacturer, according to the retailer's sales performance, can afford a rebate to them. In the literature, it is shown that the rebate can precisely influence their customers' purchase decisions. Since the retailer could pass a segment of the rebate paid by the manufacturer to the final customers, the rebate could indirectly affect consumer demand. In most research, it is generally assumed that rebates only affect the manufacturer and retailer's negligible profit rather than the consumer demand. In this chapter, demand functions that explicitly depend on the rebate paid by the firms to the end consumers are studied (J. Huang, Leng, and Parlar 2013).

The main parameters used in the functions will be described as follows:

α Rebate effectiveness (rebate rate)
β Proportion of all consumers redeem the rebate
a Potential market size
b Price elasticity
γ Rebate sensitivity
R Rebate face value
R_R Rebate provided by the retailer
R_M Rebate provided by the manufacturer
R_C Consumer rebate
w Wholesale price
c Product unit cost
ϵ Rebate independent random variable
p Retail price
D Demand for the product
δ Parameter
t Period of time
D_E Demand from consumer watching e-shop straight
D_W Product demand from consumer
θ Degree of decrease in utility watching e-shop by link

In Table 4.1 and Table 4.2, demand functions come as a single random period, including a deterministic component and a stochastic element, ϵ, defined over a finite range [A, B], with a mean of μ and a standard deviation of σ. Two functional forms of the demand error, additive and multiplicative error, are considered (Arcelus, Kumar, and Srinivasan 2008). Other types of demand functions are described respectively, and last, types of demand models on which rebates have an indirect influence on demand are explained.

Under a deterministic environment, price discrimination performs better than the manufacturer rebate. Moreover, in this environment, MIR is the best contract for the manufacturer. However, a stochastic environment leads the manufacturer rebate to perform better (Demirag et al. 2010).

TABLE 4.1
Rebate-Dependent Demand Models (Part 1)

Type		Function	Conditions and parameters	Reference
Stochastic	Additive	$D(p,R) = a + \gamma R - bp + \epsilon$	$a > 0$	(Arcelus, Kumar, and Srinivasan 2008)
	Multi-plicative	$D(p,R) = aR^{\delta} p^{-\beta} \epsilon$	$0 < \gamma < b$ $0 < \delta < 1 < \beta$ $R > 1$	(Arcelus, Kumar, and Srinivasan 2008) and (J. Huang, Leng, and Parlar 2013)
	Multi-plicative	$D(p,R) = \alpha p^{-\beta} \epsilon$	$0 < \gamma < b$ $0 < \delta < 1 < \beta$ $R \leq 1$	(Arcelus, Kumar, and Srinivasan 2008)
	Multi-plicative	$D(p,R) = g(p - \gamma R)\epsilon$	$g(0) =$ decreasing nonlinear function	(Agrawal and Smith 2010)
	Additive	$D(p, R_R R_M) =$ $a - b(p - \alpha R_R - \alpha R_M) + \epsilon$		(Geng and Mallik 2011)
	Multi-plicative	$D(R, el, \epsilon) = (\alpha R + \epsilon) el$	el : Promotional effort level	(Yang et al. 2015)
		$D(p,R) = a(p - R)^{\beta}$	$a > 0, \beta < -1,$ $R > 0$	(J. Huang, Leng, and Parlar 2013)
		$D(R): \mu(\alpha + \epsilon - p + \eta s + \gamma R)$	μ : Promotion frequency η : Green level sensitive factor s : Green level	(Lin 2020)
		$D(p, R_C) = g(p - \alpha R_C)\epsilon$	$g(0) =$ decreasing nonlinear function	(Agrawal and Smith 2010)

TABLE 4.2
Rebate-Dependent Demand Models (Part 2)

Type	Function	Conditions and parameters	Reference
Deterministic	$D(R) = a - bp + cR$		(Khouja 2006)
Conditional function	$D(R) = \dfrac{a}{b}\left(b - p + R - \dfrac{1}{2}p_0\right)$	If $R > p_0$ If $0 \leq R \leq p_0$ p_0: Consumers' value of a unit of time	(Khouja and Zhou 2010)
	$D(R) = \dfrac{a}{b}\left(b - p + \dfrac{R^2}{2p_0}\right)$		
	$D(R) = \dfrac{a}{b}(b - p + \gamma R)$		

(Continued)

TABLE 4.2

Rebate-Dependent Demand Models (Part 2)

Type	Function		Conditions and parameters	Reference
	D_E	D_W		(Zhou et al. 2017)

$$
D_E = \begin{cases} \alpha_{CW}(1-p) & b(1-\dfrac{p-\beta_{CW}p}{1-\theta}) \\[2ex] b(1-\dfrac{\beta_{CW}p}{1-\theta}) & b(\dfrac{\beta_{CW}p}{1-\theta}-\dfrac{\beta_{CW}p-p}{\theta}) \\ +\alpha_{CW}(1-p) & \\[2ex] 1-p & 0 \end{cases}
$$

Conditions:
$$\dfrac{1-\theta}{p} \le \beta_{CW} \le \alpha_{CW}$$
$$1-\theta \le \beta_{CW} \le \dfrac{1-\theta}{p}$$
$$0 \le \beta_{CW} < 1-\theta$$

CW: Cashback websites

α_{CW}: E-shop rebate rate

β_{CW}: CW rebate rate

r_E: Commission of α_{CW}

r_W: Rebate of β_{CW}

Type	Function	Conditions and parameters	Reference
Hybrid	$D(p, I(t), R_1(t), R_2(t)) =$ $\begin{cases} \alpha+\delta u_0 - bp & \text{if } u_0 \le I(t) \\ \alpha+\delta I(t) - bp + \gamma_2 R_2(t) & \text{if } u_1 \le I(t) \le u_0 \\ \alpha+\delta I(t) - bp + \gamma_2 R_2(t) & \text{if } u_1 \le I(t) \le u_0 \end{cases}$	$a>0,\ \delta>0$ u_0: Lower limit of display stock level (DSL) u_1: Upper limit of DSL I: Inventory level	(Dey et al. 2019)
Hybrid	$D_i(R_R R_M \epsilon) = y_i(R_R R_M)\epsilon \quad i \in \{e,l\}$ $y_e(R_R R_M) = a(p-\alpha R_R - \alpha R_M)^{-b}$ $y_l(R_R R_M) = a - b(p-\alpha R_R - \alpha R_M)$	$a>0,\ b>0$ e: Iso-elastic multiplicative demand function l: Linear multiplicative demand function	(Li et al. 2016)
Poisson	$\lambda_t = \begin{cases} \lambda \bar{F}(p_t) & \text{if } r_t = 0 \\ \lambda \bar{F}(p_t - \alpha(r_t - w)) & \text{if } r_t \ge w \end{cases}$	λ: Poisson process rate F: The probability of a new customer purchasing the product under a discount-only policy	(Hu, Hu, and Ye 1953)
Discount dependent	$D_t = a - bp_t - rb(p_t - p_{t-1})$	p_t: Market price in period t r: Price discount sensitivity coefficient	(Gao et al. 2017)

Type	Function	Conditions and parameters	Reference
	$\left(\rho_i + \vartheta F_R\right)D_i = \rho_i^l D_i$ $\rho_i^l = \left(\rho_i + \vartheta f_R\right)$	ρ_i: The retailer advertising effort for the item $i(\rho_i \geq 1)$ ρ_i^l: The optimal value of ρ_i f_R: Portion of cash discount ϑ: Constant parameter	(Pakhira, Maiti, and Maiti 2018)

Furthermore, a linear rebate-dependent demand function causes the optimal result simply to obtain. However, since the actual relation between demand and rebate is usually nonlinear, some attributes are missed in these models. Hence, power functions, with constant demand elasticity to the rebate, are proposed to show the characteristics of nonlinearities (J. Huang, Leng, and Parlar 2013).

4.4 CASE STUDIES AND EXAMPLES

Various forms of rebate are widely used in different industries, such as health, automobiles, hardware, software, consumer electronics, fashion, and beauty, two of which will be briefly described (Dey et al. 2019).

4.4.1 HEALTH CARE

Pharmaceutical expenditure and other medical expenditures constitute a significant amount of total expenditures on health and has increased in the last decade. In Germany and the United States, 14.8% and 11.9%, respectively, of total health expenditure consists of pharmaceutical costs (Graf 2014). One approach for reducing this cost is to create group purchasing organizations (GPOs). Research and surveys show that in the 80% of hospitals that have been examined, they reduce costs by making at least 50% of their purchases through GPOs. This process is also applied in German medical insurance. GPOs accumulate their members' demand and request an offer from the manufacturer instead of purchasing drugs and reselling them. To reduce the costs, supply contracts, which generally consist of a rebate, deal with more than one firm; therefore, the members of the GPOs could buy the policy at the prices agreed upon in the contract (Graf 2014).

Based on the conditions, three types of rebate contracts are studied: multiple, exclusive, and partially exclusive. Multiple rebate contracts decrease the overall expenditure for the members of the GPOs, and exclusive rebate contracts, unlike no rebate contracts, have no effect on the members. Therefore, both rebate contracts are beneficial. The manufacturer typically prefers multiple rebate contracts (Graf 2014).

Due to changes in drug market rules (Arzneimittelneuordnungsgesetz, AMNOG) in Germany, partially exclusive rebate contracts are often implemented (Graf 2014).

4.4.2 E-SHOP

The popularity of online shopping has grown rapidly and has become one of the most common ways of shopping. New methods, such as cashback websites (CW), were created for aiding e-shops in increasing their market share. According to the definition, provided in Wikipedia, a CW is a type of rewarding site that pays its members a percentage of money gained when they buy products and services through relevant links. Consumers make an account on the CW and from every purchase on the e-shop, the CW gets a commission. The CW then shares a fraction of the commission with consumers as rebates, adding it to the consumer's CW account so he or she can request payment to a linked bank account. The difference is given to the CW. Therefore, the cashback presented by the CW is particularly dissimilar with the usual channel and consumer rebates because, generally, the manufacturers offer the channel rebate to the retailers in regard to enhancing their sales efforts, while in the CW, the manufacturers or retailers offer the consumer rebate straight to the consumer to manipulate market demand (Zhou et al. 2017).

4.5 MATHEMATICAL MODELS

Two mathematical models in the literature that are considered rebate-dependent demand functions are elaborated hereafter.

4.5.1 HEALTHCARE SYSTEM

Based on the hoteling model (Graf 2014), considered manufacturers offer different discounts to the consumers with multiple rebate contracts. Whenever members of the GPO purchase from Manufacturer 1, it becomes more favorable than another firm as far as the rebate is concerned. These supplementary competitive advantages brought profits for Manufacturer 1, so it raises its price. Moreover, Manufacturer 2 reduces its prices to make up for lost rebates by consumers.

β	Quality differences
x	Demand for the product purchased from one of the manufacturers
R	Rebates
r	Parameter of rebate (determined by GPO, or it can be legally attributed)
tx	Linear transportation costs
p_i	Price offered by the manufacturers
D	Demand for the product

$R(x) = rx^2$ expresses the entire rebate where $r<t$. Moreover, $r<t$ and the linearity of tx cause the values in equilibrium to remain positive (Graf 2014).

Model

Two manufacturers offer all-unit discounts, so all GPO members can purchase their pharmaceutical products with the desired variety. Rebates decrease the shipping cost for each customer. The demand functions are as follows:

$$D_1^{MR}(p_1,p_2) = \begin{cases} 1 & \text{if } p_2 - p_1 \geq t - r - \beta \\ \dfrac{p_2 - p_1 + t + \beta - r}{2t - 2r} & \text{if } r - t - \beta \leq p_2 - p_1 \leq t - r - \beta \\ 0 & \text{if } p_2 - p_1 \leq r - t - \beta \end{cases} \quad (4.1)$$

And

$$D_2^{MR}(p_1,p_2) = \begin{cases} 1 & \text{if } p_1 - p_2 \geq t - r - \beta \\ \dfrac{p_1 - p_2 + t - \beta - r}{2t - 2r} & \text{if } r - t + \beta \leq p_1 - p_2 \leq t - r + \beta \\ 0 & \text{if } p_1 - p_2 \leq r - t + \beta \end{cases} \quad (4.2)$$

According to the demand functions, two manufacturers at the same time maximize their profits.

$$\pi_1^{MR} = (p_1 - c)D_1(p_1,p_2) - r\big(D_1(p_1,p_2)\big)^2 \quad (4.3)$$

$$\pi_2^{MR} = (p_2 - c)D_2(p_1,p_2) - r\big(D_2(p_1,p_2)\big)^2 \quad (4.4)$$

Solution

The two manufacturers, at the same time, make "take-it-or-leave-it" suggestions to the GPO. The GPO decreases the costs to agree with the offer made by the firm as a base price. When both firm price offers are equal, Manufacturer 1 is chosen (Graf 2014).

4.5.2 REBATES IN A TWO-ECHELON SC

Muzaffar, Malik, and Rashid (2018) considered a two-echelon SC, including a manufacturer and a retailer. The retailer purchases product from the manufacturer at a specified price and sells it to consumers at retail price. Demand is assumed to be linear and definite in terms of price and amount of rebate. Furthermore, the rebate amount should not exceed the manufacturer's marginal profit, and the retail price should not exceed the market potential. Properties will be described in the following:

Notations

W	Manufacturer price
p	Retail price
D	Demand
R	Rebate
a	Market potential
b	Price sensitivity
d	Elasticity to rebate
γ	Rebate sensitivity
$\bar{\gamma}$	Minimum fraction of rebate

Main Model

The demand function is given by $a - bp + ydR$ where $d = b = 1$. Since the retailer rebate is redeemed as soon as the purchase happens, it has a similar price reduction function for customers. This means they do not notice the distinction between price sensitivity and rebate sensitivity. Therefore, the demand function is given in the form of $a - bp + yR$. The retailer maximizes the following objective function (Muzaffar, Malik, and Rashid 2018).

$$max\pi_R = \left(p - w - \gamma R + R \right)\left(a - bp + \gamma R \right) \tag{4.5}$$

The manufacturer objective function is:

$$max\pi_M = \left(w - c - R \right)\left(a - bp + \gamma R \right) \tag{4.6}$$

s.t $0 \le R \le w - c$

Solution

To solve this model, the Stackelberg game is used in which the manufacturer, as a leader, determines the wholesale price so his profit is maximized and announces this price to the buyers/followers. This process ensures that the retailer does not change the retail price in the next level of the sales season. The retailer determines the order quantity after being informed about the rebate value and specifies the portion given to the customers with the manufacturer's consent.

It is clear that the manufacturer's performance is submodular in (γ, R). So, a reduction difference in (γ, R) is presented. For a stable pricing policy and exogenous w, P, and γ, the optimal rebate expression is as follows:

$$RR^* = \frac{1}{2}\left[\left(w - c \right) - \frac{\left(a - bp \right)}{\gamma} \right] \tag{4.7}$$

It consistently builds a win-win situation that $\left(w - c \right) > \dfrac{\left(a - bp \right)}{\gamma}$ and $\gamma \ge \overline{\gamma}$ are fulfilled at the same time, where $\overline{\gamma} \in [0,1]$ (Muzaffar, Malik, and Rashid 2018).

4.6 RESEARCH TRENDS

The basic condition of most articles examined in this chapter is based on a two-echelon SC with a newsvendor model. MIR, manufacturer rebate, retailer rebate, consumer rebate, and wholesale price rebate were the top five most common types of rebates used in the literature.

Adding an uncertain element to the demand functions leads to stochastic models, which makes them more popular than deterministic demand functions in the research. According to the competition between the manufacturer and retailer, game theory provides a win-win solution for both firms to make suitable profits, so the game theory is mostly used as a solution. The manufacturer rebate is widely used in the operations fields and has gained attention recently. First, it was used in the SC with a single manufacturer and single retailer, then developed to other kinds of SCs with different types of demand functions and solutions. Other rebate strategies, like MIR, which works as a motivational factor for manipulating the final customer's demand, are used in the additive and multiplicative stochastic demand functions. Especially in an additive stochastic framework, both the manufacturer and retailer can propose MIR to the final customers and benefit from the situation (Khouja 2006, Arcelus, Kumar, and Srinivasan 2008, Sigué 2008, Geng and Mallik 2011, Lin 2020, and Lin 2020).

Moreover, some authors consider rebates as a discount that have no direct impact on the demand function. In the stochastic functions, proper quantity discount makes benefits for all members of the SC. So, various game theory models, including cooperative and non-cooperative models, are proposed to the supplier to determine the suitable amount of discount (Ke and Bookbinder 2012).

According to the literature, different types of rebate and demand functions, the solution methods, and the SC structure are summarized in Table 4.3.

TABLE 4.3
Rebate in the SC-Related Literature

Reference	Demand type	Type of rebate	Supply chain structure	Solution
(Khouja 2006)	Deterministic	MIR	Periodic inventory model	Differential
(Arcelus, Kumar, and Srinivasan 2008)	Stochastic	Manufacturer rebate Retailer rebate	Two-echelon supply chain newsvendor model	Differential
(Khouja and Zhou 2010)	Stochastic	MIR	Single manufacture, single retailer decentralized supply chain	Game theory Stackelberg
(Agrawal and Smith 2010)	Stochastic	Manufacturer rebate Retail pricing	Newsvendor supply chain	Game theory
(Geng and Mallik 2011)	Stochastic	MIR	Single manufacturer Single retailer Supply chain	Game theory Nash equilibrium
(Valliathal and Uthayakumar 2011)	Deterministic	Price-dependent rebate	Shortage is allowed	Differential

(*Continued*)

TABLE 4.3
Rebate in the SC-Related Literature

Reference	Demand type	Type of rebate	Supply chain structure	Solution
(Saha and Sarmah 2013)	Stochastic	Instant rebate Consumer rebate	Two-echelon supply chain newsvendor model	Fuzzy
(Graf 2014)	Hybrid	Multiple exclusive Partially exclusive	Two manufacturers Competitive	Horizontal differentiation Hoteling model
(K.-L. Huang, Kuo, and Lu 2014)	Stochastic	Wholesale price rebate	Two-echelon supply chain, one manufacturer	Convexity
(Lan, Zhao, and Tang 2015)	Stochastic	Price rebate	Supply chain Principal-agent problem	Game theory
(Yang et al. 2015)	Stochastic	MIR	Newsvendor model Supply chain	Differential
(Saha and Goyal 2015)	Deterministic	Wholesale price rebate	Two-echelon supply chain	Differential
(Li et al. 2016)	Stochastic	Manufacturer rebate Retailer rebate	Two-echelon supply chain	Nash equilibrium Game theory
(Bajwa, Fontem, and Sox 2016)	Nonlinear Deterministic	Price rebate	Coordinated decision making	Lagrangian dual approach
(Nie and Du 2017)	Stochastic	Quantity discount	Dyadic supply chain One supplier, two retailers	Stackelberg
(Hu, Hu, and Ye 2017)	Deterministic	MIR Instant discount	Dynamic pricing model of limited inventory One manufacturer, one retailer	Poisson process
(Zhou et al. 2017)	Deterministic	Consumer rebate	Single manufacturer	Differential
(Gao et al. 2017)	Deterministic	Price discounts	Two-level online retail supply chain	Variance Covariance
(Muzaffar, Malik, and Rashid 2018)	Stochastic	MIR Retailer rebate	Two-level supply chain Decentralized supply chain	Stackelberg

Reference	Demand type	Type of rebate	Supply chain structure	Solution
(Ke and Bookbinder 2018)	Stochastic	Discount	Tri-level Three supply chain members (shipper, carrier, consignee) One manufacturer Multiple suppliers	Heuristic algorithm
(Pakhira, Maiti, and Maiti 2018)	Discount dependent	Cash discount Retailer rebate	Two-level supply chain Multi-item supplier-retailer	Crisp equivalent
(Zhan, Chen, and Hu 2019)	Stochastic	Trade credit sales rebate	Newsvendor model	Game theory
(Dey et al. 2019)	Hybrid	Dynamic or static rebate	One retailer Display stock level	Optimal control theory Hybrid bat algorithm
(Jazinaninejad et al. 2019)	Stochastic	Price rebate	Supply chain	Game theory
(Lin 2020)	Stochastic	Manufacturer rebate	Two-level green supply chain One manufacturer, one retailer	Stackelberg

Table 4.3 shows that most researchers consider SC with a single manufacturer and retailer, although in recent years some attention is paid to the multi suppliers or retailers. Some points can be mentioned for future research. First, develop a two-echelon SC with multi members that creates a competitive environment to make more profits. Second, use different random variables that lead to different types of demand functions and new solution approaches. Also, adding a hidden Markov chain and fuzzy equivalent for another rebate strategy, especially in the healthcare system and the online market, could cause a better solution for the problems. Third, use more rebate applications in different industries by considering multiple products instead of a single one. Finally, adding uncertain conditions to the hybrid and complicated demand functions to consider consumer probability behavior is another direction for future research.

REFERENCES

Agrawal, Narendra, and Stephen A. Smith. 2010. *Retail Supply Chain Management*. Edited by Narendra Agrawal and Stephen A. Smith. Vol. 223. Boston, MA: International Series in Operations Research & Management Science.

Arcelus, F. J., Satyendra Kumar, and G. Srinivasan. 2008. Pricing and rebate policies in the two-echelon supply chain with asymmetric information under price-dependent, stochastic demand. *International Journal of Production Economics* 113, no. 2: 598–618.

Bajwa, Naeem, Belleh Fontem, and Charles R. Sox. 2016. Optimal product pricing and lot sizing decisions for multiple products with nonlinear demands. *Journal of Management Analytics* 3, no. 1: 43–58.

Cárdenas-Barrón, Leopoldo Eduardo, Ali Akbar Shaikh, Sunil Tiwari, and Gerardo Treviño-Garza. 2020. An EOQ inventory model with nonlinear stock dependent holding cost, nonlinear stock dependent demand and trade credit. *Computers and Industrial Engineering* 139.

Chiu, Chun-Hung, Tsan-Ming Choi, Ho-Ting Yeung, and Yingxue Zhao. 2012. Sales rebate contracts in fashion supply chains. *Mathematical Problems in Engineering* 2012: 1–19.

Cho, Soo-Haeng, Kevin F. McCardle, and Christopher S. Tang. 2009. Optimal pricing and rebate strategies in a two-level supply chain. *Production and Operations Management* 18, no. 4: 426–446.

Demirag, Ozgun Caliskan, Ozgul Baysar, Pinar Keskinocak, and Julie L. Swann. 2010. The effects of customer rebates and retailer incentives on a manufacturer's profits and sales. *Naval Research Logistics (NRL)* 57, no. 1: 88–108.

Dey, Kartick, Debajyoti Chatterjee, Subrata Saha, and Ilkyeong Moon. 2019. Dynamic versus static rebates: An investigation on price, displayed stock level, and rebate-induced demand using a hybrid bat algorithm. *Annals of Operations Research* 279, no. 1–2: 187–219.

Gao, Dandan, Nengmin Wang, Zhengwen He, and Tao Jia. 2017. The bullwhip effect in an online retail supply chain: A perspective of price-sensitive demand based on the price discount in E-commerce. *IEEE Transactions on Engineering Management* 64, no. 2: 134–148.

Geng, Qin, and Suman Mallik. 2011. Joint mail-in rebate decisions in supply chains under demand uncertainty. *Production and Operations Management* 20, no. 4: 587–602.

Giri, B. C., R. Bhattacharjee, and T. Maiti. 2018. Optimal payment time in a two-echelon supply chain with price-dependent demand under trade credit financing. *International Journal of Systems Science: Operations and Logistics* 5, no. 4: 374–392.

Graf, Julia. 2014. The effects of rebate contracts on the health care system. *The European Journal of Health Economics* 15 (5): 477–487.

Heydari, Jafar, and Javad Asl-Najafi. 2021. A revised sales rebate contract with effort-dependent demand: A channel coordination approach. *International Transactions in Operational Research* 28, no. 1: 438–469.

Heydari, Jafar, Mehdi Rastegar, and Christoph H. Glock. 2017. A two-level delay in payments contract for supply chain coordination: The case of credit-dependent demand. *International Journal of Production Economics* 191: 26–36.

Hu, Shanshan, Xing Hu, and Qing Ye. 1953. Operations research. *Physics Today* 6 (3): 30–32.

Hu, Shanshan, Xing Hu, and Qing Ye. 2017. Optimal rebate strategies under dynamic pricing. *Operations Research* 65, no. 6: 1546–1561.

Huang, Jian, Mingming Leng, and Mahmut Parlar. 2013. Demand functions in decision modeling: A comprehensive survey and research directions. *Decision Sciences* 44, no. 3: 557–609.

Huang, Kwei-Long, Chia-Wei Kuo, and Ming-Lun Lu. 2014. Wholesale price rebate vs. Capacity expansion: The optimal strategy for seasonal products in a supply chain. *European Journal of Operational Research* 234, no. 1: 77–85.

Jazinaninejad, Mona, Seyed Mohammad Seyedhosseini, Seyyed-Mahdi Hosseini-Motlagh, and Mohammadreza Nematollahi. 2019. Coordinated decision-making on manufacturer's epq-based and buyer's period review inventory policies with stochastic price-sensitive demand: A credit option approach. *RAIRO—Operations Research* 53, no. 4: 1129–1154.

Ke, Ginger Y., and James H. Bookbinder. 2012. Discount pricing for a family of items: The supplier's optimal decisions. *International Journal of Production Economics* 135, no. 1: 255–264.

Ke, Ginger Y., and James H. Bookbinder. 2018. Coordinating the discount policies for retailer, wholesaler, and less-than-truckload carrier under price-sensitive demand: A tri-level optimization approach. *International Journal of Production Economics* 196: 82–100.

Khouja, Moutaz. 2006. A joint optimal pricing, rebate value, and lot sizing model. *European Journal of Operational Research* 174, no. 2: 706–723.

Khouja, Moutaz, and Jing Zhou. 2010. The effect of delayed incentives on supply chain profits and consumer surplus. *Production and Operations Management* 19, no. 2: 172–197.

Lan, Yanfei, Ruiqing Zhao, and Wansheng Tang. 2015. An inspection-based price rebate and effort contract model with incomplete information. *Computers & Industrial Engineering* 83: 264–272.

Li, Jianbin, Niu Yu, Zhixue Liu, and Lianjie Shu. 2016. Optimal rebate strategies in a two-echelon supply chain with nonlinear and linear multiplicative demands. *Journal of Industrial and Management Optimization* 12, no. 4: 1587–1611.

Lin, Zhibing. 2020. Manufacturer rebate in green supply chain with information asymmetry. *INFOR: Information Systems and Operational Research*: 1–15.

Muzaffar, Asif, Muhammad Nasir Malik, and Ammar Rashid. 2018. Rebate mechanism for the manufacturer in two-level supply chains. *Asia Pacific Management Review* 23, no. 4: 301–309.

Nie, Tengfei, and Shaofu Du. 2017. Dual-fairness supply chain with quantity discount contracts. *European Journal of Operational Research* 258, no. 2: 491–500.

Pakhira, Nilesh, Manas Kumar Maiti, and Manoranjan Maiti. 2018. Fuzzy optimization for multi-item supply chain with trade credit and two-level price discount under promotional cost sharing. *International Journal of Fuzzy Systems* 20, no. 5: 1644–1655.

Pramanik, Prasenjit, Manas Kumar Maiti, and Manoranjan Maiti. 2017. A supply chain with variable demand under three level trade credit policy. *Computers and Industrial Engineering* 106: 205–221.

Saha, S., and S. K. Goyal. 2015. Supply chain coordination contracts with inventory level and retail price dependent demand. *International Journal of Production Economics* 161: 140–152.

Saha, S., and S. P. Sarmah. 2013. Coordination of supply chains by downward direct rebate under symmetric and asymmetric information. *International Journal of Integrated Supply Management* 8, no. 4: 193.

Seifert, Daniel, Ralf W. Seifert, and Margarita Protopappa-Sieke. 2013. A review of trade credit literature: Opportunities for research in operations. *European Journal of Operational Research* 231, no. 2: 245–256.

Shaikh, Ali Akbar, Md. Al-Amin Khan, Gobinda Chandra Panda, and Ioannis Konstantaras. 2019. Price discount facility in an EOQ model for deteriorating items with stock-dependent demand and partial backlogging. *International Transactions in Operational Research* 26, no. 4: 1365–1395.

Sigué, Simon Pierre. 2008. Consumer and retailer promotions: Who is better off? *Journal of Retailing* 84, no. 4: 449–460.

Valliathal, M., and R. Uthayakumar. 2011. An EOQ model for rebate value and selling-price-dependent demand rate with shortages. *International Journal of Mathematics in Operational Research* 3, no. 1: 99.

Yang, Shuai, Ki Sung Hong, and Chulung Lee. 2014. Supply chain coordination with stock-dependent demand rate and credit incentives. *International Journal of Production Economics* 157, no. 1: 105–111.

Yang, Shilei, Charles L. Munson, Bintong Chen, and Chunming Shi. 2015. Coordinating Contracts for supply chains that market with mail-in rebates and retailer promotions. *Journal of the Operational Research Society* 66, no. 12: 2025–2036.

Zhan, Jizhou, Xiangfeng Chen, and Qiying Hu. 2019. The value of trade credit with rebate contract in a capital-constrained supply chain. *International Journal of Production Research* 57, no. 2: 379–396.

Zhao, Yingxue, Tsan-Ming Choi, T. C. E. Cheng, Suresh P. Sethi, and Shouyang Wang. 2014. Buyback contracts with price-dependent demands: Effects of demand uncertainty. *European Journal of Operational Research* 239, no. 3: 663–673.

Zhou, Yong-Wu, Bin Cao, Qinshen Tang, and Wenhui Zhou. 2017. Pricing and rebate strategies for an e-shop with a cashback website. *European Journal of Operational Research* 262, no. 1: 108–122.

5 Service Level Effects

M. Ziari and Mohsen S. Sajadieh
Department of Industrial Engineering, Amirkabir
University of Technology, Tehran, Iran

CONTENTS

5.1 INTRODUCTION

Service level can influence demand and shape it differently in distinct time horizons. Both the offline and online response systems for fulfilling customer demand need to be evaluated as the main features of service level in its correct meaning. It means they are responsible for responding to the consumer demand correctly, in an appropriate way, in an appropriate amount of time. However, service level includes many components (Baghalian et al. 2013). This description can briefly state the main concept of applying service level for business developments, promotions, customer satisfaction and other economic, social or environmental targets. In this section, we aimed to focus on the definition and objectives of service level and a description of service rate.

The performance of a system is generally measured by service level (Spreng, Harrell, and Mackoy 1995). Each system involves certain desirable goals. Service level means how we were successful in touching these goals in a supply chain or competitive market. Here, another concept different from service level emerges, known as fill rate. Many examples exist for the service level, such as the percent of answered calls in a call center, the percent of waiting customers, the percent of customers not experiencing stock-out, the percent of fulfilled demand, and so on (Craig, DeHoratius, and Ananth 2016).

To have a more obvious definition, service level equals zero if one part of an order is not filled. In other words, the service level is 51% for a case in which the whole order is fulfilled except for the items with 51% filling rate. It is usually applied in supply chains in terms of delivering to the production department. As mentioned, fill rate or service rate is different from service level and is described in the next sections.

5.1.1. SERVICE RATE

Service rate is a performing measure used to assess and evaluate the service level of customers in a market or supply chain (Shi, Song, and Powell 2013). There exist many examples for service rate, including the number of filled units in comparison with the fill rate. So, if the total number of orders is 10,000 and you can only meet 8,500 units, your fill rate is 85%.

In addition, service level is of high importance in queuing theory. Another definition for service level in queuing theory is the rate of customers serving in a system. For example, a bank client with mean service time of 2 minutes for each customer will serve an average of 30 customers in an hour (Zavanella et al. 2015).

To this aim, the effects of service level on consumer behavior and customer demand is investigated in this chapter and essential factors in shaping the demand are considered. Critical factors in service level–dependent demand are the amount of fulfilled demand, the product or service quality, lead time and system reliability (Huang et al. 2013). Hence, four types of service level–dependent, quality-dependent, lead time–dependent and reliability-dependent demand are presented to conduct practitioners. In the next section, the most prevalent service level definitions are presented. Then, service level–dependent demand in four separation modes are defined, and the model's pros and cons are described in a table. Afterward, the dependent demand functions application in supply chain management problems are analyzed to evaluate the advantages and disadvantages. Two case problems are defined before presenting optimization models considering service level effects. Afterward, a research trend and the way it was completed is reviewed and summarized in a table. Finally, the chapter concludes with the last section in which future research directions are described for future researchers.

5.2 CONCEPTUAL FRAMEWORK

In this section, the main motivating features of service level effect on customer demand will be discussed. In each problem, the most critical question is how to apply the service level concept in a problem. Therefore, different definitions of service level should be reviewed and the features of them must be understood. Here, the main factors, including type of modeling in three groups, are defined as follows.

5.2.1 EVENT-ORIENTED SERVICE LEVEL (TYPE 1)

An event-oriented performance criterion is named as α service level. This criterion assesses the probability that the total amount of the customer orders will be fulfilled by stock on hand without delay. The hint is that customers must arrive within a certain time interval (Lewis and Ray 1999).

Considering the time interval in which customers arrive, two distinct models of service level were discussed in the literature. Therefore, α means the probability of fulfilling an arbitrary order of an arriving customer by the on-hand stock with respect to a demand period. Then, period α_p service level can be written as follows:

$$\alpha_p = \text{Prob \{stock on hand } (I_0) \geq \text{ period demand\}} \tag{5.1}$$

The distribution function or probability of the stock on hand must be known for determining the safety stock, which guarantees α_p service level.

Similarly, α shows the probability of not having stock-out in an order cycle in a standard reference period. It can also be interpreted as the percentage of all order cycles without stock-outs. Then, cycle α_c service level can be written as follows:

$$\alpha_c = \text{Prob \{stock on hand } (I_0) \geq \text{ demand during replenishment lead time\}} \tag{5.2}$$

The second version is more common in operations management books and supports the idea of not facing stock shortages during the lead time (time between the reorder point and the arrival of the order). That is formulated in the following relation:

$$\text{Prob \{demand during lead time } \leq \text{ remaining amount of stock\}} \tag{5.3}$$

Notice that the reorder point is not zero or negative, the orders are incrementing one by one and the stock is monitored continuously. Thus, stock-outs are impossible before reordering.

5.2.2 QUANTITY-ORIENTED SERVICE LEVEL (TYPE 2)

Here, a quantity-oriented performance measure is introduced as β service level presenting the proportion of total demand delivering from stock on hand without delay in a reference period:

$$\beta = \frac{\text{Expected value of back-orders in a time period}}{\text{Expected demand of a period}} \tag{5.4}$$

This method is often involved with calculating a loss integral in a normal distribution function and is equal to the probability of delivering a random unit of demand on time (Lewis and Ray 1999).

Generally used in industrial cases and oppositely to the variations of α service level, the β service level influences the event of stock-out and the total back-ordered amount simultaneously. Moreover, when the zero-demand probability is 0, definitions deliver the service level $\alpha \le \beta$.

5.2.3 QUALITY- AND TIME-ORIENTED SERVICE LEVEL (TYPE 3)

Now, a quantity- and time-related performance criterion is described as γ service level (Pekgun et al. 2016). It serves to assume back-order amounts and waiting times for the total amount of back-ordered demand. The γ service level can easily be calculated as follows:

$$\gamma = 1 - \frac{\text{Expected value of back-order level in a time period}}{\text{Expected demand of a time period}} \tag{5.5}$$

The γ service level is less prevalent when applied in industrial case problems.

5.3 DEMAND FUNCTIONS

Demand denotes not only the willingness but also the ability of consumers to buy certain quantities of products, goods and services at a delivered price within a given time period (Rezapour et al. 2015). Many factors, called demand determinants, effect demand and include the production cost, its price compared to competitors' prices, its price compared to alternative or complementary products, or the consumers' levels of income (Wang, Sun, and Wang 2016). Different types of demand functions are generated in the literature to reflect customer behavior or other critical factors' effects on the demand in different markets (Grimm 2008). Service level also attracts many practitioners to focus on how it can influence and shape demand in different situations. Thus, the most common demand functions in respect to service level effects can be categorized in 4 distinct groups, including service level–dependent demand, quality-dependent demand, lead time-dependent demand and reliability-dependent demand (Huang et al. 2013). As the quality of delivered services or products, the lead time of products and the reliability of supply chains are effective in service level management, the related demand functions are discussed to aid practitioners who are applying service level concepts in their research (Choi et al. 2017). These separations and the key points of each category are briefly described in the following sections.

5.3.1 SERVICE LEVEL–DEPENDENT DEMAND

Here, studying service level–dependent demand, two distinct linear and non-linear forms are introduced. In the linear formulation, the general factors influencing demand quantity are consumers' income (M) and delivered service level (S) (Huang et al. 2013). The demand function considering the defined determinants can be written as follows:

$$D = f(M, S) \tag{5.6}$$

Now, a simple service level–dependent demand function for product Y is defined below:

$$D(S,M) = d + cM + dS \qquad (5.7)$$

Here, D is a function of the effective factors in forming the required quantity. The demand function is positively related to the income of consumers, that is, if the income is more, the demand will be more. The demand function is positively related to service level, too, that is, if the service level increases, the quantity demanded will increase.

The non-linear form of service level–dependent demand function is widely applied in the literature. It is said that the demand of consumers is determinant dependent (Federgruen and Heching 1999).

As it generally is in this field of study and presented by Kim and Lee (1998), Bernstein and Federgruen (2004), Jung and Klein (2005) and Li et al. (2016), a non-linear log-separable demand D can be written as follows:

$$D(S) = dS^{\beta} \qquad (5.8)$$

where d is basic demand, S the service level of delivering product and β is the service level coefficient. The general form can be extended by adding additional factors, including the price of products, the delivery time for each product and the rival offered price in a market.

5.3.2 Quality-Dependent Demand

Quality is the other effective factor in service level management problems (Huang et al. 2013; Xu 2013). The total demand of a final product is linearly dependent on the quality of the products.

$$D(Q) = d + \beta Q \qquad (5.9)$$

where d is the base demand, Q is the quality of products and β is the quality coefficient.

In addition, a non-linear log-separable demand D can be written as follows:

$$D(Q) = d + \beta Q \qquad (5.10)$$

where d is the base demand, Q is the quality of the products, and α is the quality coefficient.

5.3.3 Lead Time–Dependent Demand

Lead time is the other key component in service level management problems and can enhance service level easily by trivial changes (Huang et al. 2013).

The common form of linear lead time–dependent demand considering the delivering lead time can be written as follows:

$$D(L) = d - \beta L \tag{5.11}$$

where d is the base demand, L is the lead time for delivering products, and β is the lead time coefficient.

Here, two distinct non-linear forms of market lead time-dependent demand D can be introduced as follows:

$$D(L) = d_{max} - (L_{min} - L_r)\, d_k \tag{5.12}$$

where d_{max} is the maximum demand capacity, L_{min} is the provided lead time and L_r is the required lead time for each product (Altendorfer 2017). The maximum amount of available capacity less than unused capacity (in case of lacking demand) is realized by demand capacity d_k.

And the other non-linear form of lead time dependent demand function, the Cobb–Douglas model, especially for monopolist cases, can be derived as follows (Huang et al. 2013):

$$D(L) = \lambda L^{-\beta} \tag{5.13}$$

where, λ is the shape parameter in demand function, L is the lead time of the product, and β is the lead time coefficient.

5.3.4 Reliability-Dependent Demand

Reliability is the other effective factor in service level management. The total demand of the final product is linearly dependent on the reliability of a system (Huang et al. 2013).

$$D(R) = d + \beta R \tag{5.14}$$

where d is the base demand, R is the system reliability, and β is the reliability coefficient.

Here, two distinct non-linear forms of market lead time-dependent demand D can be categorized into two groups. First, the product demand based on the reliability of a product can be written as:

$$D(R) = d_0 (1 - R)^{-\alpha} \tag{5.15}$$

where the demand is an increasing function of the reliability R.

The other form of non-linear non-negative function of reliability dependent demand can be obtained as follows (Mahapatra et al. 2017):

$$D(R,t) = \alpha R^t \tag{5.16}$$

TABLE 5.1

Four General Types of Service Level–Dependent Demand Functions

Demand function	Linear	Non-linear
Service level dependent	$D(S) = d + \beta S$	$D(S) = dS^{\beta}$
Quality dependent	$D(Q) = d + \beta Q$	$D(Q) = dQ^{\alpha}$
Lead time dependent	$D(L) = d - \beta L$	$D(L) = d_{\max} - (L_{\min} - L_r)d_k$
		$D(L) = \lambda L^{-\beta}$
Reliability dependent	$D(R) = d + \beta R$	$D(R) = d_0(1 - R)^{-\alpha}$
		$D(R) = \alpha R^t$

where, α is the shape parameter in demand function and R is the reliability of product at time period t. The general dependent demand considering service level, quality, lead time and reliability is summarized in Table 5.1.

5.3.5 ADVANTAGES AND DISADVANTAGES OF DIFFERENT DEMAND FUNCTIONS

Most of the service level–dependent demand models have some general rules. In linear form, the effective factor appears with a suitable coefficient, and in non-linear form the main problem is defining the most suitable function for representing the customers' elasticity to the provided service level. Briefly, it can be said that the more the service level of a product is, the more demand for that product the sellers will suffer. Moreover, some of the advantages and drawbacks are also mentioned hereafter.

The linear form of demand function is more prevalent than the non-linear form, for some critical reasons. First, it is the simplest function that can be used as a demand model. Although, the linear form of service level–dependent demand is very straightforward, it is far from the real relations between demand and the provided service level. To be more compatible with real problems, it is more logical to define non-linear forms. Secondly, determination of the service level coefficient in the demand function is hard and cannot be easily obtained via general distribution functions. Mostly fuzzy or stochastic coefficients are the best approaches to deal with this problem.

In contrast with the general linear models, some non-linear forms, like power functions, can fit more appropriately in this case. Regarding the power model, since it can be both attained from the Cobb–Douglas production function and easily converted to a linear function, it is the most preferable one in formulating non-linear demand–service level relationships in the market. Hence, one issue must be noticed to be more realistic. There is a specific size for each market, which is not infinite, and some of these functions reflect the infinite demand. It should be solved by choosing the appropriate demand function, finding upper or lower bounds for the provided service level or by evaluating the most reasonable coefficients.

Hence, by considering a log-separable formulation as the demand function, they could not have any necessity of defining any lower and upper bounds, which can dramatically increase this function's desirability. All in all, power and linear demand functions are the most popular in the related literature.

Although it seems that the linear or non-linear demand functions are applied in the research randomly, these functions are based on the problem features and how they fit in defined cases.

Notice that the number of echelons in a supply chain can determine the best-applied function. That is, if the system is a single echelon, all functions can lead the problem to similar results, and this would provide the researchers the opportunity of selecting them randomly. However, they must pay more attention to the function selection process in multi-echelon systems since different functions result in very different conclusions. It is worth clarifying that, in the case of fixed-demand elasticity, using the power function is the smartest choice. It can be concluded that the real case structure, number of echelons, the customer elasticity and how the provided service level reflects demand (e.g., linearly or in power shape), the most suitable demand function can be selected.

5.4 SERVICE LEVEL APPLICATIONS IN DIFFERENT MODELS

The discussed models were used in different organizations and industries, including retailing, e-commerce, hospitality, manufacturing and other practical cases that are presented in this section. Mathematical models are described in the next section. Most of the service level–dependent demand functions are involved with quality and lead time as the critical factors, and the others study the formulation of utility functions, service level functions and assessing the value of reliability. Thus, most of the previous research focuses on the products and their features for maintaining or growing the service level. But it should be mentioned that the trend has recently shifted to service management. The papers or research focusing on products in service level management are in two business-to-business (B2B)— transactions between businesses—and business-to-customer (B2C)—direct transactions between businesses and customers—categories. The other research focusing service management is more involved with B2C cases. Here, some related case studies are reviewed.

5.4.1 INDUSTRIES AND EXAMPLES

Here, two real case problems are presented to find the effects of service level on the demanded quantity and are also applied to the associated demand functions in practical subjects.

To find the appropriate inventory service level, all the supply chain members must analyze the effective changes in service level shaping the demanded quantity. Substantial investments are usually required to increase inventory service level in process change and technology. Although the increase in costs from incrementing the service level can be measured, the benefits cannot be evaluated easily. (Baek and Yoon 2020). Increasing the inventory service level by a supplier not only helps capture lost sales but also changes the demanded quantity from retailers (Syntetos and Boylan 2008). Changing the inventory service level and the consequent results on the received demand from retailers in functional apparel case problems was measured in the literature.

Javed and Wu (2019) worked on the assortment of retailer products, the frequency of demand and the changes in service level as the main factors in service-level management. They focused on online retailing and attempted to evaluate the effects of this type of systems on customer satisfaction. There exist different empirical models concerning the issue of service level effects on inventory management. But a trivial number of previous studies focused on B2B settings. Craig, DeHoratius, and Ananth (2016) collaborated with Hugo Boss[1] for empirically measuring the relation between retailer demand and the service level of suppliers for functional apparel items in a supply chain. They designed the project in a way that they can properly measure the relationship among retailer orders and the provided service level of suppliers. The relationship among demand and availability of a product (as service level) was also previously investigated by empirical papers, and it was concluded that the more inventory there is, the more demand dampens.

Heim and Sinha (2001) revealed that inventory shortage can easily put customer loyalty in danger. In shortage or stock-outs, customers abandon the purchase process, switch to a different retailer or prefer substitutable products. Lieu (1991) and Anderson, Fitzsimons and Simester (2006) studied the long-term consequence of shortages and analyzed how demand changes in stock-outs.

The next application of service level management is in demand response (DR), which is generally known as the changes from normal consumption patterns in industrial usage by the customers or end-use customers in response to changes in lead time, price, availability, quality, promotions or system signals (U.S. Department of Energy). Prior research considered demand shedding for emergency case response, whereas nowadays DR projects are more targeted in large energy cases (Baek and Yoon 2020).

The other application of service level in demand can be found in electric power systems. An electric grid named the Smart Grid (SG) is responsible for delivering from generation centers to end-users in a smart and controlled way since it can change its patterns of purchasing or usage based on incentives, obtained information, and disincentives. Most of the practitioners studying service level effects confirmed that SG's capability in improving customers' responsiveness and reliability performance helps managers conveniently control the demanded quantity. Hence, demand side management (DSM), which includes all the programs required to be done on a demand section, depicts an integral and comprehensive part of SG (Shen et al. 2014). It is mostly about utility-based programs implemented for controlling the energy consumption on the demand side. Therefore, both the owners and customers benefit from DSM programs, and it can also help them operate in a better way, which will result in spot price volatility and reduce peak demand. Many case problems in industrial or service models can demonstrate the effect of service level on the demand quantity.

5.5 MATHEMATICAL MODELS

Two instances of mathematical models formulated in the literature for assessing the effect of service level in case problems are presented hereafter.

5.5.1 SERVICE LEVEL IN INVENTORY MANAGEMENT (MODEL 1)

Inventory service level demonstrates the expected percentage of not touching a unit of inventory. This probability is required to assess the buffer.[2] First, the service level describes a trade-off between the stock-out costs and inventory costs (Vermorel 2012). Now, an optimal service level is calculated by modeling the associated stock-out and inventory costs. To model the problem, the following variables should be defined:

Terminology and Notations
Parameters

P	The service level
H	The carrying cost per unit within the lead time
M	The unit cost of a stock-out

The lead time here is considered as the time scope. Therefore, $H = \dfrac{d}{365} Hy$ considering that d (described in days) is the lead time used instead of the added annual transferring cost of Hy. The optimal service level can be formulated as follows:

$$p = \Phi\left(\sqrt{2 ln\left(\frac{1}{\sqrt{2\pi}} \frac{M}{H} \right)} \right) \tag{5.17}$$

where Φ is the normal cumulative distribution function. The stock level is changing continuously, but for making the model more practical, a service level value is to be decoupled from the forecasted demand. Thus, Q is assumed as the reorder point. For a given service level, the total holding cost of the inventory and the shortage costs are combined in $C\,(p)$, which is as follows:

$$C(p) = Q(p)H + (1-p)MO \tag{5.18}$$

where $Q\,(p)\,H$ is the holding cost and MO the shortage cost occurring with (p) and $(1-p)$ percentage, respectively. As the introduced $Q\,(q)$ is the reorder point, $Q\,(p) = Z + \sigma\,\Phi^{-1}(p)$ can be used, where Z is the lead demand, σ the expected demand deviation and $\Phi(p)^{-1}$ the inverse of the normal cumulative distribution function ($\mu = 0$, $\sigma = 1$). To solve the problem, after using the proposed approach for finding out the optimal reorder point assuring the provided service level, a heuristic method can be used for finding the other optimal decision variables.

5.5.2 SERVICE LEVEL EFFECTS ON RETAILER DEMAND (MODEL 2)

For settling inventory service level, suppliers should evaluate the effects of inventory service level on demand. Here, another table of notation is presented to show another model studied by Xu, Munson and Zeng (2016).

Terminology and Notations
Parameters

ω	Fixed procurement cost
θ	Cost of procurement sensitivity to quality
τ	Discount on the fixed cost of procurement
c_b	Discount on the quality-related cost of procurement
c_e	Quality-related procurement cost that e-retailer obtains
φ	Cost sensitivity to the logistic service level
$k_e(k_t)$	Service capability of e-retailer
A	The base demand
δ	Market demand sensitivity to price
$\alpha_e(\alpha_b)$	Quality effects on market demand
$\beta_e(\beta_t)$	Logistics service effects on market demand

Decision variables

$q_e(q_b)$	Product quality under procurement mode
$s_e(s_t)$	Logistics service level
ω_b	Resale price
p	Retail price
D_{ij}	Market demand
π_b	Profit of company
π_t	Profit of the third-party logistics provider
II^{ij}	Profit of e-retailers

Now, the model can be formulated as follows:

$$Max\ \pi_b = \left(\omega_b - \tau\omega - c_b q_b^{\theta} \right) D_{bc} \tag{5.19}$$

s.t.

$$Max\ II^{ts} = \left(p - \omega_b \right) D_{bc} - k_e s_e^{\varphi} D_{bc} \tag{5.20}$$

where the market demand $D = \left(A_i - \delta p \right) q_i^{\alpha_i} s_j^{\beta_j}$ can be assessed by replacing determinants and the service level, $s_t^{DT} = (l^{DT} \dfrac{\beta t}{k_t(\varphi + \beta t)})^{\frac{1}{\varphi}}$.

After modeling the problem, three distinct logistic systems were defined, and the model was solved under three scenarios to find the optimal equilibriums in each case. Finally, the obtained results showed the best strategy for similar problems.

5.6 RESEARCH TRENDS

Determining the main trends in service level measurements has noticeable effects on improving researchers' creativity, which leads to the formation of novel extensions and contributions. Thus, with the goal of presenting the recent research trend, in this section, most related studies are proposed as different research involving service-level measurement in supply chain management problems.

Here, we aimed to focus on research streams considering service level in shaping demand functions. Hall and Porteus (2000) were among the first practitioners who built a simple dynamic model to evaluate service competition among different firms. Then, Bernstein and Federgruen (2004) proposed a general equilibrium with application in companies that were competing on price and service level. Later, Liu et al. (2007) extended Hall and Porteus (2000) and developed a general demand function by relaxing the assumption of the service failure pattern. Different from the literature, considering services in the same level of the supply chain, some focused on the service level in two- or multi-echelon supply chains for investigating the competition among supply chain members on price and service level. Chen et al. (2008) described a type of competition among manufacturers and retailers. Kurata and Nam (2010) discussed the interaction of retailer and manufacturers' after-sales services.

Moreover, some of the related research in the literature focused on quality- and service-dependent demand. In e-retailing systems, a critical service is logistics directly affecting consumers' utility. Li and Lee (1994) represented the market competition under a customer preferences model considering price, quality and delivery speed as the main components affecting service level. Hou et al. (2018) focused on e-retailing investments in delivery service, considering customers diffusion based on online reviews and purchasing experience. Gurnani and Erkoc (2008) and Ma (2013) studied a supply chain with quality- and sales service–dependent demand to form the competition among the manufacturers and retailers. Ma (2013) proved that a manufacturer will spend less money and effort on improving quality if the retailer does not exert sufficient effort on sales service. Similarly, the retailer will exert less sales-service effort if the manufacturer doesn't spend enough money and effort on improving quality. It can be said that lead time–dependent demand functions are usually implemented for analyzing inventory-related systems. In addition, few papers aimed to analyze joint optimization problems considering quality, lead time or other factors simultaneously. Quality-dependent demand functions are often used in manufacturing or service systems. But most of the previous attempts focused on the manufacturing system service level or the service quality in service systems, which can be extended by investigating the effects of product quality on the demanded quality. Moreover, researchers ignored logit function in this case, which can bring noticeable results. Finally, most of the research worked on the general function or utility-based approaches to investigate customer behavior. The other features, solution approaches and system structures in previous papers are summarized in Table 5.2 for revealing the research trend of service level concept application in supply chain problems.

5.7 CONCLUSION AND FUTURE RESEARCH DIRECTIONS

In this section, the effect of service level in different models and applications are presented to conduct the readers knowing how each model works and how they can

TABLE 5.2

Service Level in the Supply Chain–Related Literature

Reference	SC structure	Mathematical model			Solution approach
		Decision variable	**Objective function**	**Demand function**	
Bernstein and Federgruen (2004)	Multiple retailers	Price, fill rate, stock level	Profit maximization	Stochastic Linear	Nash equilibrium
Li and Lee (1994)	Multiple firms	Price, quality, delivery time		Stochastic Linear	Nash equilibrium
Hall and Porteus (2000)	Multiple firms	Number of customer service failures Service rate capacity	Capacity maximization	Linear	Nash equilibrium
Liu, Shang, and Wu (2007)	Multiple firms	Service level Order quantity	Profit maximization	Stochastic Linear	Closed-form solutions
Chen, Kaya, and Ozalp (2008)	Single manufacturer One retailer and online channel	Price, service -level, Lead time	Profit maximization	Stochastic Linear	Game theory
Kurata and Nam (2010)	Multiple firms	Service level Market share	Profit maximization	Stochastic Linear	Nash equilibrium
Hou, de Koster, and Yu (2018)	Single and multiple retailers	Cost investments	Profit maximization	Linear	Nash equilibrium
Gurnani and Erkoc (2008)	Single manufacturer, single retailer	Wholesale price, type of contract	Profit maximization	Linear	Closed form solutions
Xiao and Yang (2008)	Competing supply chains with multiple retailers	Retail price , service level	Profit maximization	Stochastic Linear	Game theory
Ma (2013)	Single manufacturer, single retailer	Wholesale price, quality effort level, marketing effort level	Profit and effort level maximization	Linear	Game theory
Mahapatra et al. (2017)	Single firm	Reliability of inventory, order quantity	Cost minimization	Non-linear	Closed form solutions

(Continued)

TABLE 5.2

Service Level in the Supply Chain–Related Literature

Reference	SC structure	Mathematical model			Solution approach
		Decision variable	Objective function	Demand function	
Xu, Tang, and Zhou (2019)	Single manufacturer, single retailer, single third-party logistics service provider	Product quality, retail price, resale price, service level	Profit maximization	Non-linear	Game theory

apply these points and techniques for real-world industrial cases. The attributes presented in prior reviews showed different effects on the overall satisfaction of customers, in comparison with consumer demand.

As Bonoma and Shapiro (1982) noted, "companies don't buy, people do." Hence, it is important to figure out the most suitable demand functions reflecting customer sensitivity to provided service level. In this chapter, the most common applied linear and non-linear service level–dependent demand functions in four distinct categories are described, which can reflect the downstream behavior. The dependent demand functions are presented as service level–dependent, quality-dependent, lead time-dependent and reliability-dependent forms.

We learn from our attempts that some practical demand functions were widely applied in different problems to build the proper relation among customers and the entities in a market. For example, linear form is mostly used for service level–, quality, lead time and reliability-dependent demand functions. The other non-linear forms, such as Cobb–Douglas or power functions, can be used in the quality- and lead time-dependent forms. Many papers involve quality- and service level–dependent demand, and the other cases are less used in the literature. Thus, to apply the most appropriate demand function based on provided service level, one must pay attention to the critical effective factors in customer behavior, number of system echelons, type of relation among the provided service level and demanded quantity and, finally, the system's structure.

To end, further research can extend in different ways. With the noticeable growth of e-commerce, e-services and Internet-based businesses, both theoretical research and empirical findings must continue to expand on previous attempts. Other important areas to be further investigated include developments in integrated systems and optimization and control systems.

It can also be interesting to incorporate different effective factors, such as lead time and quality and stocking decisions or even shelf space–allocation decisions, into the demand function.

NOTES

[1] European fashion house delivering women's and men's clothes, shoes, and accessories.
[2] Safety stock.

REFERENCES

Altendorfer, K. 2017. Relation between lead time dependent demand and capacity flexibility in a two-stage supply chain with lost sales. *International Journal of Production Economics* 194: 13–24.

Anderson, E. T., G. J. Fitzsimons, and D. Simester. 2006. Measuring and mitigating the costs of stock-outs. *Management Science* 52, no. 11: 1751–1763.

Baek, T., and S. Yoon. 2020. Looking forward, looking back: The impact of goal progress and time urgency on consumer responses to mobile reward apps. *Journal of Retailing and Consumer Services* 54: 102046.

Baghalian, A., et al. 2013. Robust supply chain network design with service level against disruptions and demand uncertainties: A real-life case. *European Journal of Operational Research* 227, no. 1: 199–215.

Bernstein, F., and A. Federgruen. 2004. A general equilibrium model for industries with price and service competition. *Operations Research* 52, no. 6: 868–886.

Bonoma, T., and B. Shapiro. 1982. *Industrial market segmentation*. Revision 2.1, 94–103. Cambridge, MA: Harvard University Graduate School of Business Administration.

Chen, K. Y., M. Kaya, and O. Ozer. 2008. Dual sales channel management with service competition. *Manufacturing & Service Operations Management* 10, no. 4: 654–675.

Choi, T. M., P. S. Chow, B. Shen, and M. L. Wan. 2017. Service quality gap analysis of fashion boutique operations: An empirical and analytical study. *IEEE Transactions on Systems, Man, and Cybernetics: Systems* 47, no. 11: 2896–2907.

Craig, N., N. DeHoratius, and Raman Ananth. 2016. The impact of supplier inventory service level on retailer demand. *Manufacturing & Service Operations Management*: 1–14.

Federgruen, A., and A. Heching. 1999. Combined pricing and inventory control under uncertainty. *Operations Research* 47, no. 3: 454–475.

Grimm, C. 2008. Evaluating baselines for demand response programs. AEIC Load Research Workshop, San Antonio, Texas.

Gurnani, H., and M. Erkoc. 2008. Supply contracts in manufacturer-retailer interactions with manufacturer-quality and retailer effort-induced demand. *Naval Research Logistics (NRL)* 55, no. 3: 200–217.

Hall, J., and E. Porteus. 2000. Customer service competition in capacitated systems. *Manufacturing & Service Operations Management* 2, no. 2: 144–165.

Heim, G. R., and K. K. Sinha. 2001. Operational drivers of customer loyalty in electronic retailing: An empirical analysis of electronic food retailers. *Manufacturing Service Operations Management* 3, no. 3: 264–271.

Hou, R., R. de Koster, and Y. Yu. 2018. Service investment for online retailers with social media, does it pay off? *Transportation Research Part E: Logistics and Transportation Review* 118: 606–628.

Huang, J., et al. 2013. Demand functions in decision modeling: A comprehensive survey and research directions. *Decision Sciences* 44, no. 3: 557–609.

Javed, M. K., and M. Wu. 2019. Effects of online retailer after delivery services on purchase intention: An empirical analysis of customer's past experience and future confidence with the retailer. *Journal of Retailing and Consumer Services* 54: 101942.

Jung, H., and C. M. Klein. 2005. Optimal inventory policies for an economic order quantity model with decreasing cost functions. *European Journal of Operational Research* 165, no. 1: 108–126.

Kim, D., and W. J. Lee. 1998. Optimal joint pricing and lot sizing with fixed and variable capacity. *European Journal of Operational Research* 109, no. 1: 212–227.

Kurata, H., and S. H. Nam. 2010. After-sales service competition in a supply chain: Optimization of customer satisfaction level or profit or both? *International Journal of Production Economics* 127, no. 1: 136–146.

Lewis, L., and P. Ray. 1999. Service level management definition, architecture, and research challenges. Seamless Interconnection for Universal Services. Global Telecommunications Conference. GLOBECOM'99. (Cat. No. 99CH37042), IEEE.

Li, L., and Y. S. Lee. 1994. Pricing and delivery-time performance in a competitive environment. *Management Science* 40, no. 5: 633–646.

Li, X., Y. Li, X. Cai, and J. Shan. 2016. Service channel choice for supply chain: Who is better off by undertaking the service? *Production and Operations Management* 25, no. 3: 516–534.

Lieu, S. 1991. Regional impacts of air quality regulation: Applying an economic model. *Contemporary Economic Policy* 9, no. 3: 24–34.

Liu, L., W. Shang, and S. Wu. 2007. Dynamic competitive newsvendors with service-sensitive demands. *Manufacturing & Service Operations Management* 9, no. 1: 84–93.

Ma, P., H. Wang, and J. Shang. 2013. Supply chain channel strategies with quality and marketing effort-dependent demand. *International Journal of Production Economics* 144, no. 2: 572–581.

Mahapatra, G. S., S. Adak, T. K. Mandal, and S. Pal. 2017. Inventory model for deteriorating items with time and reliability dependent demand and partial backorder. *International Journal of Operational Research* 29, no. 3: 344–359.

Pekgun, P., et al. 2016. Centralized vs. decentralized competition for price and lead time sensitive demand. *Decision Sciences*.

Rezapour, S., et al. 2015. Competitive closed-loop supply chain network design with price-dependent demands. *Journal of Cleaner Production* 93: 251–272.

Shen, B., G. Chatikar, Z. Lei, J. Li, G. Wikler, and P. H. Martin. 2014. The role of regulatory reforms, market changes, and technology development to make demand response a viable resource in meeting energy challenges. *Applied Energy* 130: 814–823.

Shi, N., H. Song, and W. Powell. 2013. The dynamic fleet management problem with uncertain demand and customer chosen service level. *International Journal of Production Economics* 148: 110–121.

Spreng, R. A., G. D. Harrell, and R. D. Mackoy. 1995. Service recovery: Impact on satisfaction and intention. *Journal of Services Marketing* 9: 15–23.

Syntetos, A. A., and J. E. Boylan. 2008. Demand forecasting adjustments for service-level achievement. *IMA Journal of Management Mathematics* 19, no. 2: 175–192.

Vermorel, J. 2012. Optimal service level formula for inventory optimization. https:// www.lokad.com/service-level-definition-and-formula (accessed 30 September, 2017).

Wang, Y.-Y., J. Sun, and J.-C. Wang. 2016. Equilibrium markup pricing strategies for the dominant retailers under supply chain to chain competition. *International Journal of Production Research* 54, no. 7: 2075–2092.

Xiao, T., and D. Yang. 2008. Price and service competition of supply chains with risk-averse retailers under demand uncertainty. *International Journal of Production Economics* 114, no. 1: 187–200.

Xu, D. 2013. Integrating service quality with system and information quality: An empirical test in the E-service context. *MIS Quarterly*, 777–794.

Xu, H., C. H. Munson, and S. H. Zeng. 2016. The impact of E-services on the demand of online customers. *International Journal of Production Economics* 184: 231–244.

Xu, M., W. Tang, and C. Zhou. 2019. Procurement strategies of E-retailers under different logistics distributions with quality- and service-dependent demand. *Electronic Commerce Research and Applications* 35: 100853.

Zavanella, L., et al. 2015. Energy demand in production systems: A queuing theory perspective. *International Journal of Production Economics* 170: 393–400.

6 Marketing Decisions and Efforts

Parisa Assarzadegan
Department of Industrial and System Engineering,
Isfahan University of Technology, Isfahan, Iran

Alireza Pursaeed
Department of Industrial Engineering, Sharif
University of Technology, Tehran, Iran

CONTENTS

6.1 INTRODUCTION

Marketing efforts and advertising play an important role in today's competitive age and are essential in the relationship between manufacturers, retailers, and customers. Advertising helps manufacturers and companies get to know their competitors better and win the competition with the right planning. The importance of marketing efforts and advertising can be summarized as follows: (1) introducing new business, services, and products, (2) success in business, (3) awareness of the community about the business and product, (4) demonstrating product differentiation to the customers, and (5) attracting customers.

Marketing efforts enhance product demand and profits in a supply chain (SC) and motivate consumers to buy products and, consequently, increase sales (Huang and Li 2001). Marketing efforts include activities such as advertising, marketing research, sales force, sales promotion, and sales personnel (Esmaeili, Aryanezhad, and Zeephongsekul 2009). In SCs, manufacturers and retailers can stimulate demand and differentiate their products from competitors by advertising and promotional efforts. Also, in the closed-loop supply chain (CLSC), marketing efforts to inform customers about environmental issues and remanufactured products have recently come to researchers' attention (Esmaeili, Allameh, and Tajvidi 2016; Gao et al. 2016; Li et al. 2020).

In recent years, the number of articles examining the influence of marketing efforts on SC's demand in operations management has grown significantly (Ma, Wang, and Shang 2013b; Esmaeili, Allameh, and Tajvidi 2016; Chen et al. 2017; Ma, Li, and Wang 2017; Zerang, Taleizadeh, and Razmi 2018). In operations management research, the effect of advertising and marketing efforts on the demand function has been considered in the form of linear, square root, power, and quadratic models (Esmaeili and Zeephongsekul 2010; Karray 2013; Lau, Lau, and Wang 2010; Dai and Meng 2015).

The rest of the chapter is as follows: first, a conceptual framework for advertising and marketing efforts is presented, then the list of demand functions that consider the effects of advertising and marketing efforts is shown. Next, supply chain echelons to which marketing efforts are applicable will be investigated, also some case studies and models that have used advertising and marketing efforts are studied. Then, the research trend of marketing efforts in operations management is reviewed. Finally, a conclusion and suggestions for future research are provided.

6.2 CONCEPTUAL FRAMEWORK

The details of advertising and marketing efforts are explained below.

- **Advertising**

A new definition of advertising as provided by Kerr and Richards (2020) is as follows: "Advertising is paid, owned, and earned mediated communication, activated by an identifiable brand and intent on persuading the consumer to make some cognitive, affective or behavioral change, now or in the future." Recently, in SC studies, the concept of "cooperative advertising" has grown significantly. Cooperative advertising includes national and local advertising. Manufacturers use national advertising to increases customer knowledge about a specific brand, but local advertising is used by retailers to motivate customers to buy a product. In cooperative advertising programs, manufacturers collaborate with retailers in the costs related to local advertising (Zhang et al. 2013). Aust and Buscher (2014) reviewed various cooperative advertising models in SCs.

- **Marketing Efforts**

Marketing efforts are recognized as important tools in SCs for encouraging customers to buy a product. Marketing efforts can be in the form of sales promotions, sales force, or sales personnel. Sales promotions keep the product in the customer's mind and increase demand. Some examples of promotional efforts are price discounts, coupons, offering gifts, campaigns of "buy one, get one free," and offering more with the normal price (Giri, Bardhan, and Maiti 2015). Sales promotions can also include consumer, retailer, and trade promotions. Manufacturers offer consumer promotions directly to customers. Retailers offer retailer promotions to customers. Trade promotions are those that manufacturers offer to retailers or other SC members.

Different types of trade promotions consist of display allowance, cooperative advertising, bill back, free products, and invoices (Blattberg and Neslin 1990). Green marketing (i.e., environmental marketing, sustainable marketing, and ecological marketing) includes activities such as changes in packaging, the process of creating products, and advertising (Grundey and Zaharia 2008). Marketing efforts in green SC or eco-marketing are also necessary to make customers aware of a product's green features, such as eco-friendliness or recyclability (Ding and Wang 2020).

6.3 DEMAND FUNCTIONS

As mentioned, the effect of advertising and marketing efforts on demand functions in operations management has been investigated in the forms of power, square root, linear, and non-linear models. Table 6.1 shows the list of demand functions that consider the effect of advertising (a, A) and marketing efforts (e) on the demand function. In the following demand functions, α is the market base demand and β and γ are the sensitivity of demand function to price (p) and marketing efforts. In addition to the effect of price, advertising, and marketing efforts (sales efforts and promotional efforts) on demand functions, quality efforts (θ), green degree, and collection efforts (g) are also considered. All parameters $(\alpha, \beta, b, \gamma, \lambda)$, all variables (a, A, e, p, θ, g), and all the demand functions are positive.

TABLE 6.1

List of Demand Functions and Related Parameters

Models	Authors	Demand	Variables
Advertising			
Power	(Huang and Li 2001); (Huang, Li, and Mahajan 2002);(Yue et al. 2006); (Xie and Ai 2006); (Xie and Neyret 2009); (Szmerekovsky and Zhang 2009); (Javid and Hoseinpour 2011); (Chaab and Rasti-Barzoki 2016)	$\alpha - \beta a^{-\gamma_1} A^{-\gamma_2}$	Local advertising (a) National advertising (A)
Square root	(Karray and Zaccour 2006)	$\alpha - p_n + b p_s + \gamma \sqrt{A}$	Price (p_n, p_s) Advertising (A)
	(SeyedEsfahani, Biazaran, and Gharakhani 2011); (Aust and Buscher 2012)	$\gamma_1 \sqrt{a} + \gamma_2 \sqrt{A}$	Local advertising (a) National advertising (A)
	(Xie et al. 2017)	$\rho \left(Q + \gamma \sqrt{A} \right) - \beta p_t + b p_e$	Total demand (Q) Advertising (A) Price (p_t, p_e)

(Continued)

TABLE 6.1

List of Demand Functions and Related Parameters

Models	Authors	Demand	Variables
Linear	(Zhang et al. 2013)	$\alpha(r-p)+G+\gamma_1 a+\gamma_2 A$	Reference Price (r) Price (p) Goodwill (G) Local advertising (a) National advertising (A)
	(Yang et al. 2013)	$\alpha+\gamma A$	Advertising (A)
	(Gupta, Biswas, and Kumar 2019)	$\tilde{a}-\tilde{\beta}p+\tilde{\gamma}A+\tilde{\lambda}\theta$	Price (p) Advertising efforts (A) Quality efforts (θ)
	(Yu, Wang, and Zhang 2019)	$\alpha+\gamma A+f\tau$	Advertising (A) Low-carbon emission level (τ)
	(Li and Ouyang 2016); (Li et al. 2020)	$\alpha-\beta p+\gamma A$	Price (p) Advertising (A)
	(De Giovanni 2011)	$\dot{G}=\gamma A(t)+\lambda\theta(t)-\delta G(t)$ $D=\alpha-\beta p(t)+\varepsilon G(t)$	Advertising efforts ($A(t)$) Quality efforts ($\theta(t)$) Goodwill ($G(t)$) Price ($p(t)$)
	(Ma, He, and Gu 2020)	$\gamma_1 a(t)\sqrt{1-e(t)}+\gamma_2 A(t)-\delta e(t)$	Brand advertising ($a(t)$) Generic advertising ($A(t)$) Sales efforts ($e(t)$)
	(Jørgensen, Sigué, and Zaccour 2000)	$\dot{G}=\gamma_1 A_m(t)+\gamma_2 A_r(t)-\delta G(t)$ $D=\left(k_m a_m(t)+k_r a_r(t)\right)\sqrt{G(t)}$	Long-term manufacturer's advertising ($A_m(t)$) Long-term retailer's advertising ($A_r(t)$) Short-term manufacturer's advertising ($a_m(t)$) Short-term retailer's advertising ($a_m(t)$) Goodwill ($G(t)$)
Marketing Efforts (Promotional Efforts, Sales Efforts)			
Linear	(Tsay and Agrawal 2000); (Wu 2012)	$\alpha_i-\beta p_i+b\left(p_j-p_i\right)+$ $\gamma_1 e_i-\gamma_2\left(e_j-e_i\right)$	Price (p_i, p_j) Sales efforts (e_i, e_j)
	(Taylor 2002)	γe	Sales efforts (e)
	(Tsay and Agrawal 2000); (Gurnani and Xu 2006); (Taylor 2006); (Mukhopadhyay, Su, and Ghose 2009);	$\alpha-\beta p+\gamma e$	Price (p) Sales efforts (e)

Models	Authors	Demand	Variables
	(Lau, Lau, and Wang 2010); (Dan, Xu, and Liu 2012); (Chen et al. 2017); (Ma, Li, and Wang 2017); (Zerang, Taleizadeh, and Razmi 2018); (Ke and Jiang 2020)		
	(Ma, Wang, and Shang 2013b); (Ma, Wang, and Shang 2013a); (Mondal and Giri 2020)	$\alpha - p + \gamma e + \lambda \theta$	Price (p) Marketing efforts (e) Quality efforts (θ)
	(Dash Wu 2013)	$1 - p_i + e_i + bp_j - \gamma e_j$	Price (p_i, p_j) Promotional efforts (e_i, e_j)
	(Gao et al. 2016)	$\alpha - \beta p + \gamma e + \lambda g$	Price (p) Sales efforts (e) Collection efforts (g)
	(Basiri and Heydari 2017)	$a - \beta p_1 + \gamma e_1 + \lambda \theta +$ $b_p (p_2 - p_1) - b_\theta (\theta_2 - \theta_1) -$ $b_e (e_2 - e_1)$	Price of the green product (p_1) Price of the non-green product (p_2) Environmental quality for the green product (θ_1) Environmental quality for the non-green product (θ_2) Sales efforts for the green product (e_1) Sales efforts for the non-green product (e_2)
	(Ding and Wang 2020)	$\alpha - p + \lambda \theta + \gamma e$	Price (p) Green degree (g) Promotional efforts (e)
Power	(Xing and Liu 2012)	$\alpha e^\gamma D$	Sales efforts (e) Online demand (D)
	(Esmaeili, Aryanezhad, and Zeephongsekul 2009); (Esmaeili and Zeephongsekul 2010); (Esmaeili, Allameh, and Tajvidi 2016); (Hu, Hu, and Xia 2019)	$\alpha p^{-\beta} e^\gamma$	Price (p) Marketing efforts (e)
	(Pal, Sana, and Chaudhuri 2015)	$\alpha p^{-\beta} \theta^{k\lambda} e^\gamma$	Price (p) Quality efforts (θ) Promotional efforts (e)

(Continued)

TABLE 6.1

List of Demand Functions and Related Parameters

Models	Authors	Demand	Variables
	(He et al. 2009)	$Q - \int_0^Q F\big(x|(p,e)\big)dx$	Order quantity (Q) Distribution function of demand $\big(F(x|(p,e))\big)$ Price (p) Promotional efforts (e)
Square root	(Karray 2013)	$\alpha - p + \gamma_1\sqrt{e_1} + \gamma_2\sqrt{e_2}$	Price (p) Marketing efforts (e_1, e_2)
Non-linear	(He et al. 2009)	$Q - \int_0^Q F\big(x|(p,e)\big)dx$	Order quantity (Q) Distribution function of demand $\big(F(x|(p,e))\big)$ Price (p) Promotional efforts (e)
	(Dai and Meng 2015)	$\gamma(e)\beta(p)\xi$ $\gamma(e)\big(\beta(p)+\xi\big)$	Marketing efforts (e) Price (p) Risk (ξ)
	(Chernonog, Avinadav, and Ben-Zvi 2015)	$\beta(p)\gamma(e)$	Price (p) Sales efforts (e)

6.4 APPLICATIONS OF ADVERTISING AND MARKETING EFFORTS

In this section, the effect of advertising and marketing efforts on demand functions in different supply chains, including business-to-business (B2B), business-to-customer (B2C), and closed-loop supply chain (CLSC), is examined.

- B2B advertising is a type of advertising in which companies promote their offerings (services, goods, brands, raw materials, supplies, and resources) to other businesses (e.g., governments, corporations, institutions), whereas B2C advertising is that in which businesses promote their products and services to individual people. Advertising arises in traditional media (e.g., TV, radio, newspapers, magazines) and non-traditional media (e.g., online channels, social media, mobile advertising) (Swani, Brown, and Mudambi 2020). Advertising can positively affect brand outcomes (e.g., brand loyalty, brand awareness, and brand attitudes) and financial outcomes (e.g., demands or price) (Hanssens, Wang, and Zhang 2016; De Vries, Gensler, and Leeflang 2017). Different types of marketing efforts in B2B markets include buying allowance, count and recount allowance, merchandise allowance, buy-back allowance, promotional allowance, gifts, and premiums (Fill and Fill 2005).
- Advertising and marketing efforts in the context of CLSCs that are commonly known as green advertising and green marketing could include increasing customer awareness about sustainability, green products, and environmental

concerns to increase demand and companies' profits (Hartmann and Apaolaza-Ibáñez 2012; Shen et al. 2019; Zhang, Wang, and You 2015). On the other hand, the effect of advertising in CLSCs can be for demand of both new and remanufactured products (Li et al. 2020; Li and Ouyang 2016).

6.4.1 INDUSTRIES AND EXAMPLES

In recent decades, retailers such as Best Buy, Walmart, and Carrefour have expanded their markets through marketing efforts, including advertising, attractive shelf space, sales personnel, introducing trial samples, sales promotions, and more (Chen et al. 2017). For example, Walmart has used the following promotion strategies to attract customers and increase demand:

- Advertising through social media, billboards, newspapers, and TV ads
- Offering free trials and discounts across all seasons
- Advertising in-store through the efforts of sales personnel to attract customers to try the products

Promotional efforts in the tablet market are such that Google, Asus, and Samsung customers may receive a free power bank or a cover if they buy the new Nexus 7 or Galaxy Note Pro 12.2. In the automobile market, Mercedes-Benz gives free metallic paint to customers of the E-Class Saloon model, or customers can install a seven-speed automatic gearbox to upgrade their vehicles. Another example is BMW, which has added features such as satellite navigation, Bluetooth, and voice control to the BMW 5 Series 525i/d (Tsao 2015). UPS and FedEx offer discounts on transportation costs to increase customer demand. American Eagle and A&F Clothiers devoted part of their available shelf space to special clothing for longer periods (Tsao and Sheen 2012). Burger King and McDonald's provide coupons to enhance demand (Tsao 2010). Brands like Body Shop, Ben & Jerry's, Ecover, Patagonia, Tom's of Maine, and L.L. Bean are among the pioneers in using green marketing techniques (Rivera-Camino 2007).

6.4.2 CASE STUDY

Sinha and Verma (2020) divided the benefits related to sales promotion into two categories: utilitarian benefits and hedonic benefits. Utilitarian benefits refer to those that influence the customer's evaluation through rational thought, including usefulness, convenience, and money-saving. Hedonic benefits are those that influence the customer's evaluation through intrinsic and emotional feelings, including pleasant feelings, entertainment, and value expression.

The authors examined the effects of monetary and non-monetary sales promotions on hedonic and utilitarian benefits related to sales promotions and also the effects of hedonic and utilitarian benefits on customer perceived value. They considered that the product category moderated these relations. Non-monetary sales promotions refer to incentives such as bonuses, gifts, and sweepstakes.

They studied customer perceived value for food products and personal-care products by considering monetary and non-monetary sales promotions in India. Two categories for food products, including chocolates and biscuits, and two categories for personal-care products, including soap and toothpaste, are considered. In this study, data were collected from east, west, north, and south of Madhya Pradesh (the Indian State), and 400 questionnaires were completed. Of the 400 participants, 52.3% were males and 47.7% were females; 64.5% of the participants ranged in age from 15 to 30 years, 29.8% were between 31 and 50, and 5.8% were more than 50 years old. Regarding education level, 48.2% of the respondents were graduates, 30% were under-graduates, and 21.8% had a postgraduate degree; 37.5%, 32.3%, and 30.2% of the participants were students, businessmen, and service providers, respectively. The annual income of 58.2% of the respondents was below INR 200,000, between INR 200,001 and INR 400,000 for 34%, and 7.8%, had an annual income higher than INR 400,000.

The authors applied partial least-squares–based structural equation modeling (PLS-SEM) and used the related software (Smart PLS 3). They named the category of food products Group A and the category of personal-care products Group B. The following results were obtained for Group A with 400 samples:

- The coefficient R^2 for utilitarian benefits, hedonic benefits, and customer perceived value are 0.391, 0.473, and 0.266, respectively, which shows the predictive power is moderate.
- The influence of monetary sales promotion on hedonic benefits ($p = 0.0000, t = 6.755, \beta = 0.276$) and utilitarian benefits ($p = 0.0000, t = 9.203, \beta = 0.410$) is positive and significant.
- The influence of non-monetary sales promotion on hedonic benefits ($p = 0.0000, t = 9.754, \beta = 0.489$) and utilitarian benefits ($p = 0.0000, t = 5.356, \beta = 0.290$) is positive and significant.
- The influence of hedonic benefits ($p = 0.0000, t = 8.058, \beta = 0.366$) and utilitarian benefits ($p = 0.0000, t = 4.056, \beta = 0.219$) related to sales promotion on customer perceived value is positive and significant.

The results for Group A showed that the hedonic and utilitarian benefits related to sales promotion have a positive effect on customers' perceived value.

The following results were obtained for Group B with 400 samples:

- The coefficient R^2 for utilitarian benefits, hedonic benefits, and customer perceived value are 0.330, 0.359, and 0.225, respectively, which shows the predictive power is moderate.
- The influence of monetary sales promotion on hedonic benefits ($p = 0.0000, t = 9.754, \beta = 0.307$) and utilitarian benefits ($p = 0.0000, t = 9.203, \beta = 0.410$) is positive and significant.
- The influence of non-monetary sales promotion on hedonic benefits ($p = 0.0000, t = 2.213, \beta = 0.489$) and utilitarian benefits ($p = 0.0000, t = 5.356, \beta = 0.290$) is positive and significant.

- The influence of hedonic benefits ($p = 0.024, t = 8.058, \beta = 0.366$) and utilitarian benefits ($p = 0.0000, t = 4.056, \beta = 0.219$) related to sales promotion on customer perceived value is positive and significant.

The results for Group B show that the hedonic and utilitarian benefits related to sales promotion have a positive effect on customers' perceived value.

6.5 MATHEMATICAL MODELS

In this section, two examples with related parameters and solving methods that considered the effects of advertising and marketing efforts on demand functions are presented:

6.5.1 THE FIRST EXAMPLE

Chaab and Rasti-Barzoki (2016) investigated a SC with one manufacturer and one retailer. The manufacturer produces a product at the production cost (c) and sells it to the retailer at the wholesale price (w). The retailer sells the product to customers at the retail price (p). They considered cooperative advertising between the manufacturer and the retailer.

Parameters
α Market base demand
A Sales saturate asymptote
β Demand sensitivity to price
B Demand sensitivity to advertising
γ Effectiveness of local advertising
δ Effectiveness of national advertising
c Manufacturer's production cost
d Handling cost for the retailer
v Shape parameter

Variables
w Wholesale price
p Retail price
m Retailer margin
a National advertising
q Local advertising
t Participation rate of the manufacturer

- **Demand and profit functions**

The demand function in the SC is shown in Relation (6.1):

$$D(p,a,q) = (\alpha - \beta p)^{\frac{1}{v}} \left(A - Ba^{-\gamma} q^{-\delta} \right)$$

(6.1)

In Relation (6.1), the market base demand is α, $(\alpha - \beta p)^{\frac{1}{v}}$ is the price demand function, and $\left(A - Ba^{-\gamma}q^{-\delta}\right)$ is the advertising demand function. By substituting $p = m + w$, the manufacturer and retailer's profit functions are as follows:

$$\pi_M = (w - c)(\alpha - \beta(m + w))^{\frac{1}{v}}\left(A - Ba^{-\gamma}q^{-\delta}\right) - ta - q \tag{6.2}$$

$$\pi_R = (m - d)(\alpha - \beta(m + w))^{\frac{1}{v}}\left(A - Ba^{-\gamma}q^{-\delta}\right) - (1 - t)a \tag{6.3}$$

According to Relation (6.2), the first term of the manufacturer's profit function is profits from sales of the product to the retailer, and the second and third terms are costs related to national and local advertising, respectively. The first term of the retailer's profit function (Relation 6.3) is profits from sales of the product to customers, and the second term is costs related to national advertising.

- **Solving method**

To solve the problem, they assumed the manufacturer and retailer make decisions independently and simultaneously. A Nash game is provided to obtain the equilibrium decisions of pricing, local and national advertising, and the participation rate.

To find the equilibrium points, the first derivative of Relations (6.2) and (6.3) are obtained and are equaled to zero simultaneously:

$$w^N = \frac{v}{1 + 2v} \tag{6.4}$$

$$p^N = \frac{2v}{1 + 2v} \tag{6.5}$$

$$t^N = 0 \tag{6.6}$$

$$a^N = \left(\frac{\left(\frac{\gamma}{\delta}\right)^{\delta} \gamma v}{(1 + 2v)^{\frac{1}{v} + 1}} \right)^{\frac{1}{\delta + \gamma + 1}} \tag{6.7}$$

$$q^N = \frac{\delta}{\gamma} a^N \tag{6.8}$$

6.5.2 THE SECOND EXAMPLE

Ma, Wang, and Shang (2013b) considered a SC with one manufacturer and one retailer. The manufacturer produces a product at the production cost (c) and sells it to the retailer at the wholesale price (w). The retailer sells the product to customers at the retail price (p). The retailer enhances demand through marketing efforts, and the manufacturer increases demand through quality efforts.

Parameters

α Market base demand
γ Sensitivity coefficient of demand to marketing efforts
λ Demand sensitivity to quality efforts
η Demand sensitivity to marketing efforts
ξ Profit sensitivity to quality efforts
c Production cost

Variables

w Wholesale price
p Retail price
m Retailer margin
e Marketing efforts
θ Quality efforts

- **Demand and profit functions**

The demand function in the SC is shown in Relation (6.9):

$$D(p,e,\theta) = \alpha - p + \gamma e + \lambda \theta \qquad (6.9)$$

In Relation (6.9), the market base demand is α, and demand is decreased with an increase in price (p) and increased with an increase in marketing and quality efforts (e,θ), respectively. The authors considered the retail price as $p = m + w$, in which m is the sales margin related to the retailer. So, the retailer's profit function is as follows:

$$\pi_R(p,e) = (p - w)(\alpha - p + \gamma e + \lambda \theta) - \frac{\eta e^2}{2} \qquad (6.10)$$

According to Relation (6.10), the first term of the retailer's profit function shows the profit from sales of the product to customers, and the second term is costs related to marketing efforts.

$$\pi_M(w,\theta) = (w - c)(\alpha - p + \gamma e + \lambda \theta) - \frac{\xi \theta^2}{2} \qquad (6.11)$$

According to Relation (6.11), the first term of the manufacturer's profit function shows the profit from sales of the product to the retailer, and the second term is costs related to quality efforts.

- **Solving method**

They proposed a Stackelberg game model to solve the problem; the manufacturer is the leader, and the retailer is the follower. By substituting $p = m + w$ in Relation (6.10), the retailer's profit function is as follows:

$$\pi_R(m,e) = m(\alpha - m - w + \gamma e + \lambda \theta) - \frac{\eta e^2}{2} \tag{6.12}$$

The Hessian matrix of π_R with respect to m and e is negative-definite if $2\eta > \gamma^2$, so the optimal decisions for the retailer's variables are obtained by solving the first-order derivative of Relation (6.12):

$$\frac{\partial \pi_R}{\partial m} = \alpha - 2m + \gamma e + \lambda \theta - w = 0 \tag{6.13}$$

$$\frac{\partial \pi_R}{\partial e} = \gamma - \eta e = 0 \tag{6.14}$$

By solving Relations (6.13) and (6.14), the optimal decisions for m and e are as follows:

$$m^* = \frac{\eta(\alpha + \gamma \theta - w)}{2\eta - \gamma^2} \tag{6.15}$$

$$e^* = \frac{\gamma(\alpha + \gamma \theta - w)}{2\eta - \gamma^2} \tag{6.16}$$

By substituting Relations (6.15) and (6.16), the manufacturer's profit function is as follows:

$$\pi_M(w,\theta) = \frac{\eta(w-c)(\alpha + \gamma \theta - w)}{2\eta - \gamma^2} - \frac{\xi \theta^2}{2} \tag{6.17}$$

The optimal decisions for the manufacturer's variables are obtained by solving the first-order derivative of Relation (6.17):

$$\frac{\partial \pi_M}{\partial w} = \frac{\eta(\alpha + \gamma \theta - 2w + c)}{2\eta - \gamma^2} = 0 \tag{6.18}$$

$$\frac{\partial \pi_M}{\partial \theta} = \frac{\lambda \eta(w-c)}{2\eta - \gamma^2} - \xi \theta = 0 \tag{6.19}$$

By solving Relations (6.18) and (6.19), the optimal decisions for w and θ are as follows:

$$w^* = \frac{1}{2}\left(\alpha + c + \frac{\lambda^2 \eta(\alpha - c)}{4\eta \xi - \lambda^2 - \eta - 2\lambda^2 \xi} \right) \tag{6.20}$$

$$\theta^* = \frac{\lambda\eta(\alpha-c)}{4\eta\xi - \lambda^2 - \eta - 2\lambda^2\xi} \tag{6.21}$$

6.6 RESEARCH TRENDS

In this section, articles that investigate the effect of advertising and marketing efforts on demand functions are reviewed. Table 6.2 presents research trends from 2000 to 2020 to show the effect of advertising and marketing efforts in SCs as static/dynamic models and deterministic/stochastic models. Also, the game structures that are used to solve models are stated.

- **Static/dynamic models**

In static models, the demand function and optimal decisions of advertising and marketing efforts are stationary. But, in dynamic models, time is considered in modeling and in the demand function, and optimal decisions are obtained for advertising and marketing efforts over time.

- **Deterministic/stochastic models**

In deterministic models, the demand function is deterministic and optimal decisions of advertising and marketing efforts are obtained deterministically. But, in stochastic models, the demand function is modeled stochastically and decisions of advertising and marketing efforts are obtained stochastically.

- **Game structures**

Game theory is a common solving method in operations management for decisions about marketing efforts and advertising in SCs, and a significant number of articles have used game theory to make optimal decisions regarding advertising and marketing efforts in SCs. In the reviewed articles, five types of game structures, including (1) Nash (N), (2) the Supplier-Stackelberg (SS), (3) the Manufacturer-Stackelberg (MS), (4) the Retailer-Stackelberg (RS), and (5) the cooperation between manufacturers and retailers (Co) have been identified.

- **Contract design**

The design of contracts, to coordinate SC and collaboration between manufacturers and retailers, has been the focus of researchers. The purpose of contracts, such as cooperative advertising, participation in the costs of marketing efforts, two-part tariffs, and retail price maintenance, is to increase the partnership between manufacturers and retailers to increase demand and SC's profits.

The following topics are suggested for further research: (1) studies that have examined the impact of marketing decisions over time are limited, so considering the time for advertising and marketing efforts in SC is an interesting topic in this area; (2) in most articles, the Stackelberg game approach is used to obtain decisions of advertising

TABLE 6.2

Research Trends in Advertising and Marketing Efforts

Author(s)	Model		Game Structure					Advertising/ Marketing Efforts
			N	SS	MS	RS	Co	
(Tsay and Agrawal 2000)	stat.	det.			*		*	M.E.
(Jørgensen, Sigué, and Zaccour 2000)	dyn.	det.			*		*	Adv.
(Huang and Li 2001)	stat.	det.	*		*		*	Adv.
(Huang, Li, and Mahajan 2002)	stat.	det.			*		*	Adv.
(Taylor 2002)	stat.	stoch.	*				*	M.E.
(Gurnani and Xu 2006)	stat.	det.			*		*	M.E.
(Karray and Zaccour 2006)	stat.	det.			*		*	Adv.
(Taylor 2006)	stat.	det./stoch.			*			M.E.
(Yue et al. 2006)	stat.	det.			*		*	Adv.
(Xie and Ai 2006)	stat.	det.	*		*		*	Adv.
(Esmaeili, Aryanezhad, and Zeephongsekul 2009)	stat.	det.			*		*	M.E.
(He et al. 2009)	stat.	stoch.		*			*	M.E.
(Mukhopadhyay, Su, and Ghose 2009)	stat.	det./stoch.			*		*	M.E.
(Szmerekovsky and Zhang 2009)	stat.	det.			*		*	Adv.
(Xie and Neyret 2009)	stat.	det.	*		*	*	*	Adv.
(Esmaeili and Zeephongsekul 2010)	stat.	stoch.			*	*	*	M.E.
(Lau, Lau, and Wang 2010)	stat.	det.			*			M.E.
(De Giovanni 2011)	dyn.	det.			*		*	Adv.
(SeyedEsfahani, Biazaran, and Gharakhani 2011)	stat.	det.	*		*	*	*	Adv.
(Javid and Hoseinpour 2011)	stat.	det.	*		*		*	Adv.
(Aust and Buscher 2012)	stat.	det.	*		*	*	*	Adv.
(Dan, Xu, and Liu 2012)	stat.	det.			*			M.E.
(Xing and Liu 2012)	stat.	det.			*		*	M.E.
(Wu 2012)	stat.	det.			*			M.E.
(Dash Wu 2013)	stat.	det.	*		*		*	M.E.
(Karray 2013)	stat.	det.	*		*	*	*	M.E.
(Ma, Wang, and Shang 2013b)	stat.	det.	*		*	*	*	M.E.
(Ma, Wang, and Shang 2013a)	stat.	det.			*		*	M.E.
(Zhang et al. 2013)	stat.	det.			*		*	Adv.
(Yang et al. 2013)	stat.	det.			*		*	Adv.
(Chernonog, Avinadav, and Ben-Zvi 2015)	stat.	stoch.			*			M.E.
(Dai and Meng 2015)	stat.	stoch.	*					M.E.
(Pal, Sana, and Chaudhuri 2015)	stat.	det.	*		*	*		M.E.
(Chaab and Rasti-Barzoki 2016)	stat.	det.	*		*	*	*	Adv.

Author(s)	Model		Game Structure					Advertising/ Marketing Efforts
			N	SS	MS	RS	Co	
(Esmaeili, Allameh, and Tajvidi 2016)	dyn.	det.			*			M.E.
(Gao et al. 2016)	stat.	det.	*		*	*	*	M.E.
(Li and Ouyang 2016)	stat.	det.			*			Adv.
(Basiri and Heydari 2017)	stat.	det.				*	*	M.E.
(Chen et al. 2017)	stat.	det./stoch.	*		*	*		M.E.
(Ma, Li, and Wang 2017)	stat.	det.			*			M.E.
(Xie et al. 2017)	stat.	det.			*		*	Adv.
(Zerang, Taleizadeh, and Razmi 2018)	stat.	det.			*	*		M.E.
(Gupta, Biswas, and Kumar 2019)	stat.	stoch.	*	*	*	*		Adv.
(Hu, Hu, and Xia 2019)	stat.	det.			*		*	M.E.
(Yu, Wang, and Zhang 2019)	stat.	det.			*		*	Adv.
(Ding and Wang 2020)	stat.	det.			*			M.E.
(Ke and Jiang 2020)	stat.	det.			*	*		M.E.
(Li et al. 2020)	stat.	det.			*		*	Adv.
(Ma, He, and Gu 2020)	dyn.	det,			*			Adv.
(Mondal and Giri 2020)	stat.	det.			*		*	M.E.

and marketing efforts, so articles with a dynamic game approach, such as evolutionary games and differential games in this area, are limited; (3) addressing advertising and marketing efforts for green products, remanufactured (or refurbished) products in the context of CLSC and green SC, is among the suggestions for future work; (4) the design of contracts (cooperative advertising, participation in the cost of marketing efforts, etc.) in SC to cooperate between manufacturers and retailers in decisions of advertising and marketing efforts has been considered by many researchers in this field; therefore, the design of new contracts in this area will be a topic of interest.

6.7 CONCLUSION

In this chapter, decisions of advertising and marketing efforts in operations management are studied. The demand functions that considered advertising and marketing efforts in forms of power, square root, linear, and non-linear models are reviewed. The applications of advertising and marketing efforts in different SC echelons (B2B, B2C, and CLSC) are studied. Some case studies and two mathematical models are provided. Finally, the research trend, according to advertising and marketing decisions is proposed.

REFERENCES

Aust, Gerhard, and Udo Buscher. 2012. Vertical cooperative advertising and pricing decisions in a manufacturer—retailer supply chain: A game-theoretic approach. *European Journal of Operational Research* 223, no. 2: 473–482.

Aust, Gerhard, and Udo Buscher. 2014. Cooperative advertising models in supply chain management: A review. *European Journal of Operational Research* 234, no. 1: 1–14.

Basiri, Zahra, and Jafar Heydari. 2017. A mathematical model for green supply chain coordination with substitutable products. *Journal of Cleaner Production* 145: 232–249.

Blattberg, Robert C., and Scott A. Neslin. 1990. *Sales Promotion: Concepts, Methods, and Strategies*. United States: Prentice Hall.

Chaab, Jafar, and Morteza Rasti-Barzoki. 2016. Cooperative advertising and pricing in a manufacturer-retailer supply chain with a general demand function; A game-theoretic approach. *Computers & Industrial Engineering* 99: 112–123.

Chen, Lin, Jin Peng, Zhibing Liu, and Ruiqing Zhao. 2017. Pricing and effort decisions for a supply chain with uncertain information. *International Journal of Production Research* 55, no. 1: 264–284.

Chernonog, Tatyana, Tal Avinadav, and Tal Ben-Zvi. 2015. Pricing and sales-effort investment under bi-criteria in a supply chain of virtual products involving risk. *European Journal of Operational Research* 246, no. 2: 471–475.

Dai, Jiansheng, and Weidong Meng. 2015. A risk-averse newsvendor model under marketing-dependency and price-dependency. *International Journal of Production Economics* 160: 220–229.

Dan, Bin, Guangye Xu, and Can Liu. 2012. Pricing policies in a dual-channel supply chain with retail services. *International Journal of Production Economics* 139, no. 1: 312–320.

Dash Wu, Desheng. 2013. Bargaining in supply chain with price and promotional effort dependent demand. *Mathematical and Computer Modelling* 58, no. 9: 1659–1669.

De Giovanni, Pietro. 2011. Quality improvement vs. advertising support: Which strategy works better for a manufacturer? *European Journal of Operational Research* 208, no. 2: 119–130.

De Vries, Lisette, Sonja Gensler, and Peter S. H. Leeflang. 2017. Effects of traditional advertising and social messages on brand-building metrics and customer acquisition. *Journal of Marketing* 81, no. 5: 1–15.

Ding, Junfei, and Wenbin Wang. 2020. Information sharing in a green supply chain with promotional effort. *Kybernetes* 49, no. 11: 2683–2712.

Esmaeili, Maryam, Ghazaleh Allameh, and Taraneh Tajvidi. 2016. Using game theory for analysing pricing models in closed-loop supply chain from short- and long-term perspectives. *International Journal of Production Research* 54, no. 7: 2152–2169.

Esmaeili, Maryam, Mir-Bahador Aryanezhad, and Panlop Zeephongsekul. 2009. A game theory approach in seller—buyer supply chain. *European Journal of Operational Research* 195, no. 2: 442–448.

Esmaeili, Maryam, and P. Zeephongsekul. 2010. Seller—buyer models of supply chain management with an asymmetric information structure. *International Journal of Production Economics* 123, no. 1: 146–154.

Fill, Chris, and Karen Fill. 2005. *Business-to-Business Marketing: Relationships, Systems and Communications*. United Kingdom: Pearson Education.

Gao, Juhong, Hongshuai Han, Liting Hou, and Haiyan Wang. 2016. Pricing and effort decisions in a closed-loop supply chain under different channel power structures. *Journal of Cleaner Production* 112: 2043–2057.

Giri, B. C., S. Bardhan, and T. Maiti. 2015. Coordinating a two-echelon supply chain with price and promotional effort dependent demand. *International Journal of Operational Research* 23, no. 2: 181–199.

Grundey, Dainora, and Rodica Milena Zaharia. 2008. Sustainable incentives in marketing and strategic greening: The cases of Lithuania and Romania. *Technological and Economic Development of Economy* 14, no. 2: 130–143.

Gupta, Rohit, Indranil Biswas, and Sushil Kumar. 2019. Pricing decisions for three-echelon supply chain with advertising and quality effort-dependent fuzzy demand. *International Journal of Production Research* 57, no. 9: 2715–2731.

Gurnani, Haresh, and Yi Xu. 2006. Resale price maintenance contracts with retailer sales effort: Effect of flexibility and competition. *Naval Research Logistics (NRL)* 53, no. 5: 448–463.

Hanssens, Dominique M., Fang Wang, and Xiao-Ping Zhang. 2016. Performance growth and opportunistic marketing spending. *International Journal of Research in Marketing* 33, no. 4: 711–724.

Hartmann, Patrick, and Vanessa Apaolaza-Ibáñez. 2012. Consumer attitude and purchase intention toward green energy brands: The roles of psychological benefits and environmental concern. *Journal of Business Research* 65, no. 9: 1254–1263.

He, Yong, Xuan Zhao, Lindu Zhao, and Ju He. 2009. Coordinating a supply chain with effort and price dependent stochastic demand. *Applied Mathematical Modelling* 33, no. 6: 2777–2790.

Hu, Jing, Qiying Hu, and Yusen Xia. 2019. Who should invest in cost reduction in supply chains? *International Journal of Production Economics* 207: 1–18.

Huang, Zhimin, and Susan X. Li. 2001. Co-op advertising models in manufacturer—retailer supply chains: A game theory approach. *European Journal of Operational Research* 135, no. 3: 527–544.

Huang, Zhimin, Susan X. Li, and Vijay Mahajan. 2002. An analysis of manufacturer-retailer supply chain coordination in cooperative advertising. *Decision Sciences* 33, no. 3: 469–494.

Javid, A. Ahmadi, and Pooya Hoseinpour. 2011. A game-theoretic analysis for coordinating cooperative advertising in a supply chain. *Journal of Optimization Theory and Applications* 149, no. 1: 138–150.

Jørgensen, Steffen, Simon Pierre Sigué, and Georges Zaccour. 2000. Dynamic cooperative advertising in a channel. *Journal of Retailing* 76, no. 1: 71–92.

Karray, Salma. 2013. Periodicity of pricing and marketing efforts in a distribution channel. *European Journal of Operational Research* 228, no. 3: 635–647.

Karray, Salma, and Georges Zaccour. 2006. Could co-op advertising be a manufacturer's counterstrategy to store brands? *Journal of Business Research* 59, no. 9: 1008–1015.

Ke, Hua, and Ying Jiang. 2020. Equilibrium analysis of marketing strategies in supply chain with marketing efforts induced demand considering free riding. *Soft Computing* 25, no. 3: 2103–2114.

Kerr, Gayle, and Jef Richards. 2020. Redefining advertising in research and practice. *International Journal of Advertising*: 1–24.

Lau, Amy Hing Ling, Hon-Shiang Lau, and Jian-Cai Wang. 2010. Usefulness of resale price maintenance under different levels of sales-effort cost and system-parameter uncertainties. *European Journal of Operational Research* 203, no. 2: 513–525.

Li, Qiubin, Hui Sun, Hao Zhang, Wei Li, and Mi Ouyang. 2020. Design investment and advertising decisions in direct-sales closed-loop supply chains. *Journal of Cleaner Production* 250: 119552.

Li, Wei, and Mi Ouyang. 2016. Advertising decisions of new and remanufactured products under direct sales model. *Kybernetes* 45, no. 9: 1452–1471.

Ma, Peng, Kevin W. Li, and Zhou-Jing Wang. 2017. Pricing decisions in closed-loop supply chains with marketing effort and fairness concerns. *International Journal of Production Research* 55, no. 22: 6710–6731.

Ma, Peng, Haiyan Wang, and Jennifer Shang. 2013a. Contract design for two-stage supply chain coordination: Integrating manufacturer-quality and retailer-marketing efforts. *International Journal of Production Economics* 146, no. 2: 745–755.

Ma, Peng, Haiyan Wang, and Jennifer Shang. 2013b. Supply chain channel strategies with quality and marketing effort-dependent demand. *International Journal of Production Economics* 144, no. 2: 572–581.

Ma, Shigui, Yong He, and Ran Gu. 2020. Dynamic generic and brand advertising decisions under supply disruption. *International Journal of Production Research*: 1–25.

Mondal, Chirantan, and Bibhas C. Giri. 2020. Pricing and used product collection strategies in a two-period closed-loop supply chain under greening level and effort dependent demand. *Journal of Cleaner Production* 265: 121335.

Mukhopadhyay, Samar K., Xuemei Su, and Sanjoy Ghose. 2009. Motivating retail marketing effort: Optimal contract design. *Production and Operations Management* 18, no. 2: 197–211.

Pal, Brojeswar, Shib Sankar Sana, and Kripasindhu Chaudhuri. 2015. Two-echelon manufacturer—retailer supply chain strategies with price, quality, and promotional effort sensitive demand. *International Transactions in Operational Research* 22, no. 6: 1071–1095.

Rivera-Camino, Jaime. 2007. Re-evaluating green marketing strategy: A stakeholder perspective. *European Journal of Marketing* 41, no. 11: 1328–1358.

SeyedEsfahani, Mir Mehdi, Maryam Biazaran, and Mohsen Gharakhani. 2011. A game theoretic approach to coordinate pricing and vertical co-op advertising in manufacturer—retailer supply chains. *European Journal of Operational Research* 211, no. 2: 263–273.

Shen, Bin, Shenyan Liu, Ting Zhang, and Tsan-Ming Choi. 2019. Optimal advertising and pricing for new green products in the circular economy. *Journal of Cleaner Production* 233: 314–327.

Sinha, Somesh Kumar, and Priyanka Verma. 2020. Impact of sales promotion's benefits on perceived value: Does product category moderate the results? *Journal of Retailing and Consumer Services* 52: 101887.

Swani, Kunal, Brian P. Brown, and Susan M. Mudambi. 2020. The untapped potential of B2B advertising: A literature review and future agenda. *Industrial Marketing Management* 89: 581–593.

Szmerekovsky, Joseph G., and Jiang Zhang. 2009. Pricing and two-tier advertising with one manufacturer and one retailer. *European Journal of Operational Research* 192, no. 3: 904–917.

Taylor, Terry A. 2002. Supply chain coordination under channel rebates with sales effort effects. *Management Science* 48, no. 8: 992–1007.

Taylor, Terry A. 2006. Sale timing in a supply chain: When to sell to the retailer. *Manufacturing & Service Operations Management* 8, no. 1: 23–42.

Tsao, Yu-Chung. 2010. Managing multi-echelon multi-item channels with trade allowances under credit period. *International Journal of Production Economics* 127, no. 2: 226–237.

Tsao, Yu-Chung. 2015. Cooperative promotion under demand uncertainty. *International Journal of Production Economics* 167: 45–49.

Tsao, Yu-Chung, and Gwo-Ji Sheen. 2012. Effects of promotion cost sharing policy with the sales learning curve on supply chain coordination. *Computers & Operations Research* 39, no. 8: 1872–1878.

Tsay, Andy A, and Narendra Agrawal. 2000. Channel dynamics under price and service competition. *Manufacturing & Service Operations Management* 2, no. 4: 372–391.

Wu, Cheng-Han. 2012. Price and service competition between new and remanufactured products in a two-echelon supply chain. *International Journal of Production Economics* 140, no. 1: 496–507.

Xie, JiaPing, Ling Liang, LuHao Liu, and Petros Ieromonachou. 2017. Coordination contracts of dual-channel with cooperation advertising in closed-loop supply chains. *International Journal of Production Economics* 183: 528–538.

Xie, Jinxing, and Song Ai. 2006. A note on Cooperative advertising, game theory and manufacturer—retailer supply chains. *Omega* 34, no. 5: 501–504.

Xie, Jinxing, and Alexandre Neyret. 2009. Co-op advertising and pricing models in manufacturer—retailer supply chains. *Computers & Industrial Engineering* 56, no. 4: 1375–1385.

Xing, Dahai, and Tieming Liu. 2012. Sales effort free riding and coordination with price match and channel rebate. *European Journal of Operational Research* 219, no. 2: 264–271.

Yang, Jing, Jinxing Xie, Xiaoxue Deng, and Huachun Xiong. 2013. Cooperative advertising in a distribution channel with fairness concerns. *European Journal of Operational Research* 227, no. 2: 401–407.

Yu, Chao, Chuanxu Wang, and Suyong Zhang. 2019. Advertising cooperation of dual-channel low-carbon supply chain based on cost-sharing. *Kybernetes* 49, no. 4: 1169–1195.

Yue, Jinfeng, Jill Austin, Min-Chiang Wang, and Zhimin Huang. 2006. Coordination of cooperative advertising in a two-level supply chain when manufacturer offers discount. *European Journal of Operational Research* 168, no. 1: 65–85.

Zerang, Emad Sane, Ata Allah Taleizadeh, and Jafar Razmi. 2018. Analytical comparisons in a three-echelon closed-loop supply chain with price and marketing effort-dependent demand: Game theory approaches. *Environment, development and sustainability* 20, no. 1: 451–478.

Zhang, Juan, Qinglong Gou, Liang Liang, and Zhimin Huang. 2013. Supply chain coordination through cooperative advertising with reference price effect. *Omega* 41, no. 2: 345–353.

Zhang, Linghong, Jingguo Wang, and Jianxin You. 2015. Consumer environmental awareness and channel coordination with two substitutable products. *European Journal of Operational Research* 241, no. 1: 63–73.

7 A Company's Reputation

Mojdeh Younesi and Maryam Esmaeili
Department of Industrial Engineering, Faculty of
Engineering, Alzahra University, Tehran, Iran

CONTENTS

7.1 INTRODUCTION

In today's business environment, companies strive to build favorable relationships with their stakeholders. Nowadays, reputation is an essential tool for achieving a sustainable competitive advantage and maintaining a long-term relationship with stakeholders, including customers. Reputation is the most unique intangible property that helps a company continue its operations. Corporate reputation is about understanding customers, providers, personnel, and industry groups, and the identity of a company is made by the actors inside and outside an organization (Kirkwood and Gray 2009). Thus, how the company satisfies the demand of customers has a significant impact on its future. Therefore, this chapter empirically starts with the conceptual framework of the proposed factors. This is supported by a list of demand functions, essential variables, advantages, and disadvantages of those functions. The next part presents industries and products affected by reputation, and it is continued with relevant case studies, an example of an optimization model, and research trends. Finally, the results and conclusion close the chapter.

7.2 CONCEPTUAL FRAMEWORK OF THE PROPOSED FACTORS

Considering the erratic nature of today's economy, there is a battle for success between organizations (Nguyen 2020), whether one organization succeeds or not depends on many elements, one of which is reputation. A company's reputation is a perception of its stakeholders and is considered a competitive advantage because of its ability to influence the performance of the company. Therefore, this factor helps the corporation achieve its goals and objectives. A company's reputation is defined as a stakeholder's perception of a corporation's ability to create value relative to competitors (Rindova et al. 2005). The reputation is a series of beliefs about a company's capability in keeping its customers and stakeholders satisfied and happy.

The reputation is defined as an estimate of the consistency of an entity's attribute, which means a company can be globally renowned for each of its features, such as price, the quality of its goods, management, and creativity (Herbig and Milewicz 1993). A company's reputation might fail in fulfilling every customer's expectations. The company might be incapable of having the features that were mentioned earlier all at once. Reputation is created over time, a build-up of judgments by customers. The actual reputation reveals itself when the company has worked with different groups.

Economics, organizational theory, and marketing researchers have studied corporate reputation. Economists study matters of reputation, which is connected to the quality of goods and prices (Shapiro 1983). A company's reputation reflects the effects that a company has on its substantial customers and stakeholders. It is the overall judgment of the customers that builds a reputation for a company.

The reputation is essential for the success of companies, along with customer satisfaction and product quality. The company's reputation can be considered a mirror of its history that serves to provide information about the quality of its services or products compared to its competitors with similar target groups (Yoon, Guffey, and Kijewski 1993). The nature of the company's reputation depends on the outcomes of perceived actions. If the company repeatedly fulfills its promises, it must have a good reputation. Conversely, a company's failure to respect its stated goals may create a negative reputation (Nguyen and Leblanc 2001).

Most early works that have influenced the studies of management come from economics researchers' use of game theory to see how past interactions of individuals affect critical businesses in the future (Shapiro 1983). Researchers see reputation as assumptions about a company's type of strategy, for example, the quality of production or the ability to withstand competition. It is defined as signals that are based on a company's actions over time or under certain situations. These actions reveal fundamental indiscernible information that indicates the features of a company, multiple features that result in various incentives, or strength to take specific steps (Barnett and Pollock 2012). Hence, to achieve a comprehensive definition of reputation, Barnett, Jermier, and Lafferty (2006) categorized the inventory of purposes into three distinct clusters:

1. Reputation as a state of understanding: A company's reputation is an aggregation of perceptions and representation of knowledge because it reflects the awareness of the company.
2. Reputation as an assessment: A company's reputation is a criterion, judgment, or evaluation.
3. Reputation as an asset: A company's reputation is defined as an intangible source of economic or financial investment.

The company's reputation can be based on various sources that rely on direct experience, including products, services, and sales, or on indirect sources to shape their perceptions of the company's reputation. The better the knowledge of the customer is, the more the company's reputation develops, and that is why interacting with the customers and knowing their opinions on the products matter. Customers can also help the company build its brand just by talking about their products and services. Thus, creating a positive image for the company and its brand is done by bringing more value to customer purchases and meeting their expectations (Nguyen 2020).

The company's reputation is affected by many factors. Noted by the Reputation Institute, elements such as the company's ethical behavior, social impression, and justice positively affect its reputation (Reputation Institute 2018). Other features like the company's creativity in making products, the quality of its goods and services, a vivid image of the company's future, providing competitive benefits, transparency in business, and the capability of providing for customers' needs also have an impact on a company's reputation. The reputation of the company itself is another element that alters reputation. For instance, if a general manager has an unfavorable reputation, then the whole company's reputation is in danger of ignominy. Things like purchasing habits, levels of the economy, culture, communication, and location of a company are also serious influencers of reputation (Zraková, Demjanovičová, and Kubina 2019).

Economists analyze reputation issues concerning the product quality and price (Shapiro 1983). The reputation is a crucial factor in public relations activities, many types of research state that organizations with a good reputation have a clear advantage, as it can attract and keep loyal customers. Therefore, not only does reputation affect customer loyalty but it also inclines customers to recommend a brand to others (Shapiro 1983; Walsh et al. 2009). Another work that hints at the effect of corporate reputation on customer loyalty (Loureiro and Kastenholz 2011), done on RHEs (rural hospitality enterprises), shows factors such as satisfaction and the reputation context of RHEs. Satisfied RHE customers were more likely to buy, repurchase, and recommend to other RHE customers as well (Peña, Jamilena, and Molina 2013). In a broader sense, reputation also changes the perception of a company's attractiveness to various stakeholders, including potential employees, investors, and customers. The company's social performance also plays a role in these organizational results. Research shows that just as corporate social performance modifies reputation, it also directly affects the attractiveness of an organization to customers, investors, and potential employees (Martin, CBE, and Burke 2012).

As argued by Wang, Lo, and Hui (2003), reputation plays an even more prominent figure in services, as the evaluation of the quality of the services, especially in repurchases, is quite vague. Moreover, in services and products in which the assessment of quality by the customer is not possible or not accessible, reputation is of extreme importance. Thus, organizations may benefit more from their reputation than the actual quality of the services (Fombrun 1996). A good reputation improves the customers' trust toward one's services and advertisements and decreases cognitive dissonance since reputation acts as a stand-in for information. Through maintaining steady customer retention, companies receive higher purchase rates and achieve price premiums. Nevertheless, any organization with a good reputation could be in shambles tomorrow. The reputation of a company shows customers' usual expectations when they choose to use a company's services. It is both a strategical and financial property, not merely a motive for better marketing and an essential matter for companies that offer services to customers to gain their loyalty (Nguyen 2020). This connection between customer satisfaction and loyalty is seen in both online and real-world environments (Walsh and Beatty 2007). Besides, the reliability of estimating one's reputation can be something personal, thus much more reliable when considering customers' past experiences with a company (Nguyen and Leblanc 2001).

The companies that own a good reputation add to the price of their goods or services since customers are willing to pay for a product that is branded rather than just an ordinary product. Research on service marketing shows the relevance of intermediary influence of variables like satisfaction and loyalty, which effects attitudes in the outcome (Walsh, Dinnie, and Wiedmann 2006; Wang, Lo, and Hui 2003). The effects of intermediary influencers (satisfaction and commitment) are of particular importance. In their understanding, there is a causal relationship between these variables and influencers. Understanding how customer loyalty, commitment, and corporate reputation work together in creating different kinds of CCBs (customer citizenship behaviors) gives insight into the drivers of possible customer attitudes. Five aspects of reputation can be seen in their CBR (customer-based reputation) scale: quality of management and employer, customer orientation, financial stability and strength, quality of products and services, and social responsibility.

Numerous studies have examined the benefits of a positive company's reputation for firms. Smith, Smith, and Wang (2010) stated that according to signaling theory, a company's reputation can be considered an awareness sign about the behavior of the company and the quality of its performance. If a company manages this, it will be a strategic driver of an organization's reputation and will lead to success. This is the difference between well-known companies and those who fail (Bartikowski and Walsh 2011).

Because of this reputation, the uncertainty decreases, the company's reputation is defined as a precious asset, and firms do care about reputation in the economic framework. Researchers investigate that a company's reputation is also associated with improving the financial results and value of the company. A significant number of works that have been done on reputation shows that "good" characteristics are essential in boosting a company's value by resulting in a more convenient deal (Walsh and Beatty 2007). The reputation assists companies in raising their sales and market stocks (Shapiro 1982). Companies will be able to maintain a connection with the customers who are loyal to them (Nguyen and Leblanc 2001). In today's environment, which has proven to be extremely competitive, it is agreed that corporate reputation changes customer loyalty. A good reputation is a precious feature that leads a company toward having positive effects on various groups of stakeholders. Therefore, many researchers focus on the relationship between financial performance and the reputation of a company. The methodology of such research can be questionable due to their mixed results (Yungwook 2001). While a report by Rose and Thomsen (2004) states that good financial performance builds a firm's reputation, another report by Sabate and Puente (2003) says there is a two-way relationship between performance and reputation. As the reports vary in their results, the relationship between them is generally seen as a positive one, whether in profits or market value.

Many marketing researchers found corporate reputation to play a vital role in customer behavior. This is vital when companies want to gain customers' trust. Things like the quality of the product, the reputation of a company, and brand image have a powerful impact on the relationship between the loyalty and satisfaction of one's customers (Nguyen and Leblanc 2001). The determinant of a company's competitive performance is reputation, alongside interactions across marketing and R&D (Dutta, Narasimhan, and Rajiv 1999).

Many kinds of research indicate that a company's reputation influences the company's performance positively. On the other side, some studies show opposite results. They claim that reputation results from fair market performance (Martin, CBE, and Burke 2012). The reputation of a company has a strong effect on every single feature of a company's performance; the awareness of an investor, company equity, demands of the customer, maintaining loyal customers, and staff attraction (Highhouse, Brooks, and Gregarus 2009). Many elements play an essential role in the evolution of a company's reputation, just like how reputation modifies its outcome and performance. Looking through the financial records of a company shows that the reputation of one company is noticeably connected to its financial performance (Duhé 2009).

In summary, a positive company's reputation brings many valuable financial, competitive, and strategic benefits that enhance a firm's performance.

7.3 LIST OF DEMAND FUNCTIONS

A variety of mathematical models have been investigated to characterize demand functions, which depend on reputations. Researchers in business and economics are increasingly using such demand functions. This survey includes demand functions by considering essential variables, its marketing applications, and case studies. In addition, Table 7.1 considers where these demand functions are applied in supply chain echelons.

7.4 INDUSTRIES AND PRODUCTS AFFECTED BY THE REPUTATION FACTOR

Measuring the reputation and relevant criteria to determine the reputation of any company is essential. In the following, a summary of the studies conducted on the impact of the company's reputation on stakeholders, especially customer needs, are presented by focusing on the methodology and its results.

7.5 RELEVANT CASE STUDIES

7.5.1 Specialty Coffee Auction

The growing role of quality production in the global coffee market is recorded well. Teuber and Herrmann (2012) focused on the effect of regional origin on market prices. Since the specialty coffee market is analyzed, their model depicts some of the key features of the specialty coffee market. Because the reputation and objective

TABLE 7.1

Demand Functions

Author(s)	Main function	Variables	Functions' advantages and disadvantages	B2B or B2C marketing	Case studies
Khmelnitsky and Singer (2009)	$r_t = \gamma r_{t-1} + (1-\gamma)(1 - P(x_t))$ $d_t = r_{t-1} d$	r_t : Corporate reputation $P(x_t)$: Fraction of potential; customers who are not satisfied with the corporation's performance d_t : A product demand γ : A constant rate	It includes the system's reputation as a major factor influencing the future demand.	B2C	Numerical example
Teuber and Herrmann (2012)	$q_i^D = a + b.p_i$ $+ c.SCORE_i$ $+ d.ORIGIN_i$ $+ e.SCORE_i.ORIGIN_i$ $+ f.Z_i$	q_i^D : Demand $SCORE_i$: Sensory quality $d.ORIGIN_i$: Direct effect of reputation on demand Z_i : Other product characteristics		B2C	Data from eight coffee-producing countries from 2003 to 2009
Khmelnitsky and Singer (2015)	$r_{t+1} = \gamma r_t + (1-\gamma)G(x_{t+1})$ $D_t = r_t D$	r_t : Retailer's reputation D_t : A product demand γ : Customer behavior	The demand probability distribution is determined by the retailer's reputation at that period.	B2B	Numerical example

Author(s)	Main function	Variables	Functions' advantages and disadvantages	B2B or B2C marketing	Case studies
Li et al. (2016)	$D_r = \rho a - b_1 p_r + b_2 p_m + \beta_r \theta$	D_r : Demand for green products in the retail channel ρ : Degree of customer loyalty a : Basic market demand p : Price θ : Green degree of products	The dual-channel green supply chain depends on customer loyalty to the retail channel and cost satisfaction.	B2B/B2C	Numerical example
Albaladejo, González-Martínez, and Martínez-García (2016)	$T_t = \beta_0 + \beta_1 T_{t-1} + \gamma'.X_t + \varepsilon_t$	T_t : Current demand β_0, β_1 : The estimated coefficient that indicates the reputation X_t : Explanatory variables (income, price, etc.)	The reputation effect is affected by the ratio of lagged demand/ carrying capacity.	B2C	Spanish tourism regions in 2000–2013
Khosroshahi, Rasti-Barzoki, and Hejazi (2019)	$\text{CSI} = \int_0^\mu D d\mu$	CSI: The customer satisfaction index, which is related to CSR (corporate social responsibility) and sustainability. D: The demand function of a product μ : Transparency degree	It considers the confidence of customers about corporate activities and products, the pleasure of the whole society, and the positive relationship between CSR and demand.	B2B/B2C	Automobile company

(Continued)

TABLE 7.1
Demand Functions

Author(s)	Main function	Variables	Functions' advantages and disadvantages	B2B or B2C marketing	Case studies
Seyedhosseini et al. (2019)	$D_i\left(CSR, P_{r_i}, P_{r_j}\right)$ $= \alpha_i - \theta\exp(-CSR)$ $P_{r_i} + \gamma\exp(-CSR)\,P_{r_j}$	CSR: Manufacturer's corporate social responsibility P_{r_i} : Unit price of retailer i α_i : Market size of retailer i θ : Self-price coefficient of demand γ : Cross-price coefficient of demand		B2B	Numerical example
Housein et al. (2019)	$CSR_{it} = \beta_0 CSC_{it} + \beta_1 controlVariables_{i,t}$ $+\gamma_k \sum\limits_{n=1}^{7} controlVariables_{i,t}$ $+\sum\limits_{n=1}^{13} Year + \sum\limits_{N=1}^{59} INDUSTRY + \vartheta_{i,t}$	CSR_{it} : Corporate social responsibility performance of firm i in year t CSC_{it} : Customer demand for social capital of firm i in year t	This function improves CSR to obtain demands by customers from countries with high social capital.	B2B	Non-financial US-listed firms in 2000–2013
Yunus (2003)	$q_t = p_t^{-\eta} A_t^{\omega} f\left(R_t\right)$	q_t : Total demand at time t p: Price A : Advertising-like expenditure R : Current reputation	Since reputation is perceived as information about previous activity, it is assumed that the time derivative	B2B	-

Author(s)	Main function	Variables	Functions' advantages and disadvantages	B2B or B2C marketing	Case studies
			is proportional to the demand value of the additional oral carriers.		
Hosseini-Motlagh et al. (2019)	$D_T = R_0 + \tau x$	D_T : Total demand, which is the sum of the retail and online demand functions R_0 : Initial supply of the returned products τ : The sensitivity to the collector's CSR investment x : The collector's investment in CSR efforts		B2C	A remanufacturer company
Wang et al. (2019)	$Q_m(t) = \alpha - \beta P(t) + \eta G(t)$	$Q_m(t)$: A market demand α : A market capacity for the product β : Price sensitivity of the product η : Sensitivity of the advertisement $G(t)$: The reputation of the manufacturer $P(t)$: Retail price of the product	There is a positive correlation between demand and reputation, and when the price reduces the demand increases.	B2B	Numerical example

(Continued)

TABLE 7.1
Demand Functions

Author(s)	Main function	Variables	Functions' advantages and disadvantages	B2B or B2C marketing	Case studies
Zheng and Ren (2019)	$Auditor_{i,t} = \alpha_0 + \alpha_1 CSR_{i,t} + \alpha_2 Size_{i,t}$ $+ \alpha_3 Lev_{i,t} + \alpha_4 Growth_{i,t}$ $+ \alpha_5 Rev_{i,t} + \alpha_6 Inv_{i,t}$ $+ \alpha_7 Curr_{i,t} + \alpha_8 Roa_{i,t}$ $+ \alpha_9 Loss_{i,t} + \alpha_{10} Age_{i,t}$ $+ \alpha_{11} Top1_{i,t} + \alpha_{12} Seo_{i,t}$ $+ \alpha_{13} Soe_{i,t} + Year$ $+ Industry + \varepsilon_{i,t}$	$Auditor$: Audit demand i: Index firm t: Index year CSR: CSR information	This function studies the effect of voluntary CSR and its effect on audit demand.	B2B	Firms on the Shanghai and Shenzhen Stock Exchanges
Bruna and Nicolò (2020)	$Ri_{t+1} = \sum_{j=1}^{N} \alpha_j Fi_t\left(s_j\right)$	Ri_{t+1} : The reputation of a company i at time $t+1$ α_j : Influence of stakeholders in building the reputation, which depends on stakeholders' demands. $Fi_t\left(s_j\right)$: The trust invigorated by each member j of the N stakeholders toward the enterprise i at time t	This mathematical formalization matches the stakeholders' expectations through corporate social commitment (CSC), which includes the relationship	B2B/B2C	Numerical example

Author(s)	Main function	Variables	Functions' advantages and disadvantages	B2B or B2C marketing	Case studies
			between the young companies' five-year survival rate and the corporate reputation.		-
Fang and Huang (2020)	$d_g^t = 1 - \dfrac{B^t}{V\left(p_{gh}^t - p_{nh}^t\right)}$	d_g^t : Demand for workers with good reputations B^t : Market bonus for good reputation workers V : Upper bound of customer utility for high-quality service p_{gh}^t : Fraction of good-reputation workers serving high quality p_{nh}^t : Fraction of normal-reputation workers serving high quality	It links subsidization to reputation in a high-quality service platform.	B2C	

quality of the product are vital for increasing demand, they collected all available data for Cup of Excellence auctions over a period of six years in eight coffee-producing countries (Costa Rica, Bolivia, El Salvador, Colombia, Guatemala, Honduras, Brazil, and Nicaragua) and analyzed the e-auction markets. Therefore, a reduced form hedonic pricing model is proposed to explain auction prices. The results show that it is crucial to recognize quality, reputation, and their interactions as factors determining the price of the coffee auction and the effects of different origins in market segments. Their model is formulated based on three functions as supply function, demand function that is related to quality, regional origin, product characteristic and direct effect of reputation on demand. The results show that significant coefficients of the country-of-origin variables indicate that superior reputation leads to price rewards in the specialty coffee market. These outcomes are significant for coffee producers who decide to enter new consumer markets with their coffee.

7.5.2 The Impact of an Unsustainable Reputation on the Tourism Demand Model

Albaladejo, González-Martínez, and Martínez-García (2016) concentrated on reputation and persistence effects to support their impact on tourism demand. Their research determined a relationship between lagged and current demand has a non-fixed reputation effect. Their estimation is based on panel data from 17 Spanish regions over 13 years, and it depends on whether the tourists are domestic or international. The results show that reputation is not fixed in both estimates. Their model assumes a reputation effect that enhances demand, which is between the current one and the lagged one. This demand model includes a quadratic form of lagged demand, and a reputation effect is not fixed but is related to congestion.

7.6 EXAMPLES OF OPTIMIZATION MODELS

This example is adapted from Seyedhosseini et al. (2019). In this model, a social demand price sensitivity for a two-echelon supply chain (SC) has been proposed. In this SC, the manufacturer invests in CSR efforts and retailers compete on selling price. Thus, the main contribution of their model is to propose a new price-dependent competitive demand in which price sensitivity and customer price depend on the efforts of the manufacturer's CSR. The profit function of retailer i contains revenue minus the purchase cost of the products. To investigate the supply chain, the notations are presented as CSR (manufacturer's corporate social responsibility effort), p_{r_i} (the unit price of retailer i), γ (cross-price coefficient of demand), α_i (market size of retailer i), β (CSR effort cost coefficient of the manufacturer), p_m (manufacturer's unit wholesale price), θ (self-price coefficient of demand), c (manufacturer's unit production cost).

Hence, the profit function of retailer i ($TP_{r_i}(p_{r_i})$) and manufacturer m ($TP_m(CSR)$) are calculated as follows:

$$TP_{r_i}(p_{r_i}) = (p_{r_i} - p_m)(\alpha_i - \theta \exp(-CSR)p_{r_i} + \gamma \exp(-CSR)p_{r_i} \qquad (7.1)$$

$$TP_m(CSR) = (p_m - c)\left(\alpha_1 + \alpha_2 + (\gamma - \theta)\left(p_{r_1} + p_{r_2}\right)\exp(-CSR)\right) - \beta CSR \quad (7.2)$$

In the following, the Stackelberg and Nash game between two echelons are modeled, and the Cournot and Collusion behaviors of two retailers are examined, too. Therefore, this structure considers four games: Nash-Collusion, Stackelberg-collusion, Stackelberg-Cournot, and Nash-Cournot. The readers can find the comparison of models and the results of running the coordination scheme in all four structures in the mentioned paper. This model considers the effect of CSR made by members on the supply chain, customer price sensitivity, coordination, and the competition mechanism. The most important result in this paper reveals that the Nash-Collusion game leads to the most profits in which the maximum level of the manufacturer is CSR.

7.7 RESEARCH TRENDS

To prepare this chapter, all sources related to organizational reputation and company reputation from 1990 to 2020 are examined. To create a suitable framework, the general concepts related to reputation are studied and then, with a more detailed study, the importance and benefits of reputation are established. On the other hand, according to the main approach of the book and to investigate the effect of reputation factors on the demand function, a more in-depth study was performed on articles to identify these functions, which in terms of application and practical examples are provided in Table 7.1. Since reputation measurement is one of the most critical actions of organizations, a complete review is performed to identify the measurement methods and their results in various industries, discussed in Table 7.2. Finally, two case studies are examined in general, and an example of an optimization model with a game theory approach is presented.

TABLE 7.2

Sample Industries and Their Methods for Considering Reputation

Author(s)	Sample and industry	Method(s)	Result(s)
Nguyen and Leblanc (2001)	• 171 clients of service industries • 222 customers in the retail sector • 395 freshmen in educational services	• Tipping analyze • Chi-squared test	• The more favorable the image and reputation of the company, the higher the customer's loyalty.
Roberts and Dowling (2002)	• 15 years of companies contained in *Fortune*'s Most Admired	• Proportional hazards regression • Autoregressive profit	• A good reputation has the excellent financial performance. • A firm with a financial reputation has a strong effect on profit persistence.

(Continued)

TABLE 7.2
Sample Industries and Their Methods for Considering Reputation

Author(s)	Sample and industry	Method(s)	Result(s)
Aqueveque and Ravasi (2006)	• 4 consumption of goods and service companies	• Structural Equation Models (SEM)	• Reputation has positive effects on the perceptions of customers and services.
Chuang, Chen, and Liou (2009)	• 52 customers in international airlines in Taiwan	• Fuzzy AHP method	• Morality and service are critical to the airline image and customer demand. • Criterion makes better judgments, reduces the gap between actual performance and the desired level, and meets customer expectations.
Helm (2007)	• 1,120 individual German investors in an international customer goods	• Standardized questionnaire • Partial least squares (PLS)	• The reputation has a direct impact on significant loyalty and an indirect effect on behavioral loyalty.
Lloyd (2007)	Customers of: • Architecture construction • IT • Tourism • Health products • Waste management	• Standard questionnaire • Multiple regression analysis • Correlation coefficient tests • SEM	• Nine factors lead to the introduction of a reputation, such as financial performance, products, and brand. • Different stakeholder groups have different ratings of these factors and evaluate a company's reputation diversely.
Whelan and Davies (2007)	• 800 shoppers of retailer stores	• Standard questionnaire • 5-point Likert scale	• Agreeableness is the most critical feature of a company's reputation that will increase customer satisfaction. • Competence is the second most crucial factor in customer satisfaction.
Keh and Xie (2009)	• 351 customers of three Chinese companies in different industries of B2B services	• Standard questionnaire • 7-point scale • SEM • Maximum probability estimation method	• The company's reputation has a positive effect on customer trust and customer recognition. • Customer identification and commitment are closely related but are distinct structures in B2B regulation.
Zhang (2009)	• A total of 1,000 customers of China Mobile, BMW, Haier Group, and Siemens	• Standardized questionnaire • PLS • SEM	• Likeability has a greater and more significant impact on customer loyalty. • The company's performance and social responsibility are the most critical factors affecting the company's reputation and play a role in creating customer loyalty.

Author(s)	Sample and industry	Method(s)	Result(s)
Zaim et al. (2010)	• 280 customers of Telekom Company	• Standardized questionnaire • SEM • 5-point measurement scale	• The company reputation has a significant impact on customer satisfaction. • The reputation is based on the company's innovation, reliability, contribution, and how to satisfy customers.
Kim (2010)	• 642 customers of UK Retailers	• Standardized questionnaire • PLS • SEM	• Five variables of service quality, role-based reputation, personality-based reputation, customer satisfaction, and customer commitment are considered. • The quality of services and the emotional aspect of corporate reputation has a better role in determining customer satisfaction and loyalty.
Ozturk, Cop, and Sani (2010)	• Managers of the hotel industry	• Standardized questionnaire • Correlation coefficient tests	• Quality services, customer focus, and identifying the needs and requirements of customers for increased reputation. • Companies with a high quality of customer service benefit from skilled and innovative employees who increase customer satisfaction.
Ali, Alvi, and Ali (2012)	• 250 customers of the cellular industry	• SPSS and AMOS • Correlation analysis • Regression analysis • Reliability analysis • Fit index analysis	• There is a positive relationship between reputation, customer satisfaction, and loyalty. • There are positive effects of customer satisfaction on loyalty.
Tuck (2012)	• 24 customers of the Australian mining industry	• Delphi method	• The results indicate the existence of specific stakeholders, stakeholder network effects, and reputation dependence in the industry.
Petrokaitė and Stravinskienė (2013)	• 87 customers of chain restaurants	• Standardized questionnaire • Kolmogorov–Smirnov test	• All factors of customer reputation, namely, the quality of goods and services, emotional attractions, vision, leadership, and social responsibility, should be used more in restaurant communication activities.

(Continued)

TABLE 7.2

Sample Industries and Their Methods for Considering Reputation

Author(s)	Sample and industry	Method(s)	Result(s)
Aguwa, Olya, and Monplaisir (2017)	• 2,775 customers of the auto industry	• Association rule learning to build a customer satisfaction index (CSI) • Dynamic critical to quality (CTQ) method	• This model helps companies reduce costs, have fewer complaints, and increase customer loyalty, which is the main factor in reputation measurement.
Castilla-Polo et al. (2018)	• 76 cooperatives in the food Industry	• SEM • PLS	• The companies with a high reputation in cooperatives will have superior performances. • Certification, awards, innovations, and social responsibility are essential factors in cooperatives.

7.8 CONCLUSION AND FUTURE RESEARCH DIRECTIONS

Today, more than ever, it is critical to examine the impact of a company's reputation on the performance and needs of stakeholders, including customers. This is significant because public awareness of the company's actions has increased and the demand of individuals and organizations for greater transparency has gone higher. On the other hand, the customer's experience of the company's products and services and more customer communication through various media has caused the importance of evaluating the company's reputation to be given more attention.

As a result, the concept of company reputation along with its importance and benefits are proposed, and the appropriateness of multiple methodologies, such as survey, methods, and results, is discussed. The essential points extracted from reviewing the surveys performed are as follows.

Variables such as capability, compared to other variables such as competence, have a significant effect on customer loyalty and increase customer demand. In addition, the company's performance and social responsibility are two critical factors affecting the company's reputation and play an essential role in creating loyalty.

The company's reputation also plays a moderating role in the relationship between customer satisfaction and loyalty, as well as satisfaction and recommendation to use the service and improve upon it. Customer satisfaction will also increase the reputation of service companies.

The company's reputation has a positive effect on both customer trust and customer identity. Customer commitment mediates the relationship between customer trust and identity. Therefore, customer identity and commitment are closely related.

The components of a company's reputation include innovation, leadership, performance, customers, and products and services. Accordingly, corporate reputation dimensions include the quality of management, innovation, value, ability to attract and retain customers, responsiveness to the community, the environment, and the extensive use of company assets.

On the other hand, the relationship between corporate reputation and profitability may affect the relationship between reputation and customer loyalty, and reputation has a vital role in the customer satisfaction model.

Furthermore, the company's reputation, being at the top of the news and social media, affect the company's profitability. Highly reputable companies have superior financial performance and lower capital costs, too. Besides, corporate reputation can be a critical source of competitive advantage. One way to create the best performance of a company is to have an excellent corporate reputation. Social responsibility, quality of services, and financial performance are essential factors in gaining and maintaining reputation.

Finally, the company's reputation is created through the perception of stakeholders, especially customers, and is a kind of evaluation of the investment. Companies provided high-quality services to customers, have high financial strength, more innovation and value for customer satisfaction, and have the best position in the market.

An interesting extension would be to include the relationship between corporate image, corporate reputation, customer loyalty, and customer demand. There is a lack of evidence on the relationship between corporate image and corporate reputation in industries.

Future research should examine the relationship between the company's reputations with other stakeholder groups' needs and the firm's financial performance dynamics. The future studies can be extended to a larger supply chain by considering different stakeholders for common demands.

Finally, it will be challenging to consider demand functions stochastic by focusing on reputations in the long run.

REFERENCES

Aguwa, C., M. H. Olya, and L. Monplaisir. 2017. Modeling of fuzzy-based voice of customer for business decision analytics. *Knowledge-Based Systems* 125: 136–145.

Albaladejo, I. P., M. I. González-Martínez, and M. P. Martínez-García. 2016. Nonconstant reputation effect in a dynamic tourism demand model for Spain. *Tourism Management* 53: 132–139.

Ali, I., A. K. Alvi, and R. R. Ali. 2012. Corporate reputation, consumer satisfaction and loyalty. *Romanian Review of Social Sciences*, no. 3.

Aqueveque, C., and D. Ravasi. 2006. Corporate reputation, affect, and trustworthiness: An explanation for the reputation-performance relationship. 10th Annual Corporate Reputation Institute Conference, New York.

Barnett, M. L., J. M. Jermier, and B. A. Lafferty. 2006. Corporate reputation: The definitional landscape. *Corporate Reputation Review* 9, no. 1: 26–38.

Barnett, M. L., and T. G. Pollock. 2012. *The Oxford Handbook of Corporate Reputation*. Oxford, United Kingdom: Oxford University Press.

Bartikowski, B., and G. Walsh. 2011. Investigating mediators between corporate reputation and customer citizenship behaviors. *Journal of Business Research* 64, no. 1: 39–44.

Bruna, M. G., and D. Nicolò. 2020. Corporate reputation and social sustainability in the early stages of start-ups: A theoretical model to match stakeholders' expectations through corporate social commitment. *Finance Research Letters* 101508.

Castilla-Polo, F., D. Gallardo-Vázquez, M. I. Sánchez-Hernández, and M. C. Ruiz-Rodríguez. 2018. An empirical approach to analyse the reputation-performance linkage in agri-food cooperatives. *Journal of Cleaner Production* 195: 163–175.

Chuang, M. L., W. M. Chen, and J. J. H. Liou, 2009. A fuzzy MCDM approach for evaluating corporate image and reputation in the airline market. *The International Symposium on the Analytic Hierarchy Process.* 29 July–1 August. Pittsburgh, PA. 1–15.

Duhé, S. C. 2009. Good management, sound finances, and social responsibility: Two decades of US corporate insider perspectives on reputation and the bottom line. *Public Relations Review* 35, no. 1: 77–78.

Dutta, S., O. Narasimhan, and S. Rajiv. 1999. Success in high-technology markets: Is marketing capability critical? *Marketing Science* 18, no. 4: 547–568.

Fang, Z., and J. Huang. 2020. When reputation meets subsidy: How to build high quality on demand service platforms. IEEE Infocom 2020-IEEE Conference on Computer Communications, IEEE, Toronto, Canada. 944–953.

Fombrun, C. J. 1996. *Reputation: Realizing Value from the Corporate Image.* Boston, MA: Harvard Business School Press.

Helm, S. 2007. The role of corporate reputation in determining investor satisfaction and loyalty. *Corporate Reputation Review* 10, no. 1: 22–37.

Herbig, P., and J. Milewicz. 1993. The relationship of reputation and credibility to brand success. *Journal of Consumer Marketing.*

Highhouse, S., M. E. Brooks, and G. Gregarus. 2009. An organizational impression management perspective on the formation of corporate reputations. *Journal of Management* 35, no. 6: 1481–1493.

Hosseini-Motlagh, S. M., M. Nouri-Harzvili, T. M. Choi, and S. Ebrahimi. 2019. Reverse supply chain systems optimization with dual channel and demand disruptions: Sustainability, CSR investment and pricing coordination. *Information Sciences* 503: 606–634.

Housein, S., I. El Kalak, W. S. Leung, and Q. Wang. 2019. Why do firms do good? A customer-demand perspective. The Third Conference on CSR, the Economy and Financial Markets, Düsseldorf, Germany.

Keh, H. T., and Y. Xie. 2009. Corporate reputation and customer behavioral intentions: The roles of trust, identification and commitment. *Industrial Marketing Management* 38, no. 7: 732–742.

Khmelnitsky, E., and G. Singer. 2009. A stochastic inventory control problem with reputation-dependent demand. *IFAC Proceedings Volumes* 42, no. 4: 1689–1693.

Khmelnitsky, E., and G. Singer. 2015. An optimal inventory management problem with reputation-dependent demand. *Annals of Operations Research* 231, no. 1: 305–316.

Khosroshahi, H., M. Rasti-Barzoki, and S. R. Hejazi. 2019. A game theoretic approach for pricing decisions considering CSR and a new consumer satisfaction index using transparency-dependent demand in sustainable supply chains. *Journal of Cleaner Production* 208: 1065–1080.

Kim, J. 2010. The link between service quality, corporate reputation and customer responses. A thesis of Degree of Doctor of Philosophy University of Manchester, Faculty of Humanitiesm, United Kingdom.

Kirkwood, J., and B. Gray. 2009. from entrepreneur to mayor: Assessing the impact of the founder is changing reputation on Hubbard Foods Ltd. *Australasian Marketing Journal (AMJ)* 17, no. 2: 115–124.

Li, B., M. Zhu, Y. Jiang, and Z. Li. 2016. Pricing policies of a competitive dual-channel green supply chain. *Journal of Cleaner Production* 112: 2029–2042.

Lloyd, S. 2007. Corporate reputation: Ontology and measurement. Doctoral dissertation, Auckland University of Technology, School of Business, New Zealand.

Loureiro, S. M. C., and E. Kastenholz. 2011. Corporate reputation, satisfaction, delight, and loyalty towards rural lodging units in Portugal. *International Journal of Hospitality Management* 30, no. 3: 575–583.

Martin, M. G., C. L. C. CBE, and R. J. Burke. 2012. *Corporate Reputation: Managing Opportunities and Threats*. England: Gower Publishing, Ltd.

Nguyen, D. 2020. *Relationship Marketing as a Tool to Improve Business Activities at Helsinki Bakery*. Finland: Laurea University of Applied Sciences.

Nguyen, N., and G. Leblanc. 2001. Corporate image and corporate reputation in customers' retention decisions in services. *Journal of Retailing and Consumer Services* 8, no. 4: 227–236.

Ozturk, Y., S. Cop, and R. A. Sani. 2010. *The Effect of Corporate Reputation Management as a Competition Tool on Tourism Businesses*. Caesars Hospitality Research Summit, Las Vegas: University of Nevada.

Peña, A. I. P., D. M. F. Jamilena, and M. Á. R. Molina. 2013. Antecedents of loyalty toward rural hospitality enterprises: The moderating effect of the customer's previous experience. *International Journal of Hospitality Management* 34: 127–137.

Petrokaitė, K., and J. Stravinskienė. 2013. Corporate reputation management decisions: Customer's perspective. *Inžinerinė ekonomika*: 496–506.

Reputation Institute. 2018. www.reptrak.com/.

Rindova, V. P., I. O. Williamson, A. P. Petkova, and J. M. Sever. 2005. Being good or being known: An empirical examination of the dimensions, antecedents, and consequences of organizational reputation. *Academy of Management Journal* 48, no. 6: 1033–1049.

Roberts, P. W., and G. R. Dowling. 2002. Corporate reputation and sustained superior financial performance. *Strategic Management Journal* 23, no. 12: 1077–1093.

Rose, C., and S. Thomsen. 2004. The impact of corporate reputation on performance: Some danish evidence. *European Management Journal* 22, no. 2: 201–210.

Sabate, J. M. F., and E. Q. Puente. 2003. In practice empirical analysis of the relationship between corporate reputation and financial performance: A survey of the literature. *Corporate Reputation Review* 6: 161–177.

Seyedhosseini, S. M., S. M. Hosseini-Motlagh, M. Johari, and M. Jazinaninejad. 2019. Social price-sensitivity of demand for competitive supply chain coordination. *Computers and Industrial Engineering* 135: 1103–1126.

Shapiro, C. 1982. Consumer information, product quality, and seller reputation. *The Bell Journal of Economics*: 20–35.

Shapiro, C. 1983. Premiums for high quality products as returns to reputations. *The Quarterly Journal of Economics* 98, no. 4: 659–679.

Smith, K. T., M. Smith, and K. Wang. 2010. Does brand management of corporate reputation translate into higher market value? *Journal of Strategic Marketing* 18, no. 3: 201–221.

Teuber, R., and R. Herrmann. 2012. towards a differentiated modeling of origin effects in hedonic analysis: An application to auction prices of specialty coffee. *Food Policy* 37, no. 6: 732–740.

Tuck, J. 2012. *A Stakeholder Model of Reputation: The Australian Mining Industry*. The Business School Working Paper Series: 002-2012, Business School University of Ballarat, Victoria, Australia.

Walsh, G. K., and S. E. Beatty. 2007. Customer-based corporate reputation of a service firm: Scale development and validation. *Journal of the Academy of Marketing Science* 35, no. 1.

Walsh, G. K. Dinnie, and K.-P. Wiedmann. 2006. How do corporate reputation and customer satisfaction impact customer defection? A study of private energy customers in Germany. *Journal of Services Marketing* 20, no. 6: 412–420.

Walsh, G. K., V. W. Mitchell, P. R. Jackson, and S. E. Beatty. 2009. Examining the antecedents and consequences of corporate reputation: A customer perspective. *British Journal of Management* 20, no. 2: 187–203.

Wang, M., Y. Li, M. Li, L. Wan, L. Miao, and X. Wang. 2019. A comparative study on recycling amount and rate of used products under different regulatory scenarios. *Journal of Cleaner Production* 235: 1153–1169.

Wang, P., Y. H. Lo, and Y. V. Hui. 2003. The antecedents of service quality and product quality and their influence on bank reputation: Evidence from the banking industry in China. *Managing Service Quality* 13: 72–83

Whelan, S., and G. Davies. 2007. A comparative study of the corporate reputation of Irish versus British retailers: Lessons for retail band practice. *Irish Marketing Review* 19, no. 1/2: 37.

Yoon, E., H. J. Guffey, and V. Kijewski. 1993. The effects of information and company reputation on intentions to buy a business service. *Journal of Business Research* 27, no. 3: 215–228.

Yungwook Kim PhD. 2001. The impact of brand equity and the company's reputation on revenues. *Journal of Promotion Management* 6: 89–111.

Yunus, M. 2003. Reputation versus supplier-induced demand in the market of medical care: An application of optimal control model. https://www.semanticscholar.org/paper/REPUTATION-VERSUS-SUPPLIER-INDUCED-DEMAND-IN-THE-OF-Yunus/3b295dec6a6bcb518f149959db6f755234661a30, http://citeseerx.ist.psu.edu/viewdoc/download?doi=10.1.1.199.8910&rep=rep1&type=pdf.

Zaim, S., A. Turkyilmaz, M. Tarim, B. Ucar, and O. Akkas. 2010. Measuring customer satisfaction in Turk Telekom company using structural equation modeling technique. *Journal of Global Strategic Management* 7: 89–99.

Zhang, Y. 2009. A study of corporate reputation's influence on customer loyalty based on PLS-SEM model. *International Business Research* 2, no. 3: 28.

Zheng, P., and C. Ren. 2019. Voluntary CSR disclosure, institutional environment, and independent audit demand. *China Journal of Accounting Research* 12, no. 4: 357–377.

Zraková, D., M. Demjanovičová, and M. Kubina. 2019. Online reputation in the transport and logistics field. *Transportation Research Procedia* 40: 1231–1237.

8 Congestion in the System

Ata Jalili Marand
Department of Economics and Business Economics,
Aarhus University, Aarhus, Denmark

Seyyed Saber Mousavi Gargari
Faculty of Engineering and Natural Sciences,
Sabanci University, Istanbul, Turkey

CONTENTS

8.1 INTRODUCTION

This chapter reviews different approaches to the customer choice behavior modeling in congested systems from the operations management viewpoint. Congested systems are characterized by the inherent delay between customer order placement and the service/product delivery. This inevitable delay can be found in both service and manufacturing environments (e.g., call centers, restaurants, consultancy services, and make-to-order production systems) and is a result of product customization and/or value co-creation. Customers tend to find delays undesirable, and hence delay is among the main factors influencing customers' choice behavior in congested systems. Understanding how customers perceive the value of a service/product in the presence of delay is the key to customer choice modeling in congested systems. This forms a tight connection between the operations management literature and the literature on the psychology of waiting, which strives to explain how human beings perceive delay (see Allon and Kremer 2018 for a recent review).

Congestion is the product of stochasticity (e.g., in customer inter-arrival times or service/processing times) and limited resources (e.g., processing capacity or available products). These characteristics can be found in all of the examples previously mentioned. Stochasticity and limited resources are inherent in these examples, and the two factors bring about the formation of queues. This is the reason why queueing theory has been the preferred tool for modeling the operations of such systems.

According to Hassin (2016), the queueing theory research can be categorized based on the number of decision-makers involved in the study as: (i) performance analysis, (ii) optimal design, and (iii) analysis of choice behavior. In the first category, there is no decision-maker. The main goal of the studies in this category is to derive the performance measures of the queueing systems under different assumptions regarding the customers' inter-arrival times, service times, number of servers, queue regime, etc. Research in the second category assumes that a central decision-maker designs the system. The differences between the first and second category are that some of the parameters are subject to decision and that these decisions are made with explicit economic considerations (see, e.g., Stidham 2009). In the last category, there are at least two decision-makers, that is, the service provider(s) and the customer(s). This chapter falls within the last category.

Interactions among multiple decision-makers can be investigated by means of game theory, which would require behavior modeling at the individual level. Therefore, it is common to use models of individual customer choice rather than models of aggregate demand in the queueing literature with multiple decision-makers (see, e.g., Talluri and Van Ryzin 2004, Chapter 7, for more details on different approaches for demand modeling). In other words, the mainstream belief is that the aggregate behavior of the system can be better understood through individual-level modeling of choice behavior. The queuing literature uses utility functions to capture each individual customer's valuation of the service/products. In this chapter, we aim at providing a review of different factors considered to influence customer choice behavior in congested systems.

8.2 CUSTOMER CHOICE BEHAVIOR

In this section, we model customer choice behavior and show how the aggregate arrival rate (demand) can be derived from individual customer decisions. For the ease of exposition, we use a simple single-server queue to model the operations of the system under consideration. However, the central idea behind customer behavior modeling and customer equilibrium behavior holds for more complex queueing systems. The system under consideration can be either a service system or a manufacturing system. To be consistent throughout the rest of the chapter, however, we will refer to the system as a *service system* that is controlled by a *service provider* who provides a certain type of *service* for the customers. The system specifications are described in the following.

Customers arrive to the system following a Poisson process with rate Λ. Their service requirements are independently and identically distributed and follow an exponential distribution with rate μ. μ is referred to as the service rate. The operations of the system can therefore be modeled as an $M/M/1$ queue. The system is assumed to be operating under the first-come-first-serve discipline.[1]

Upon her arrival, a customer needs to decide on joining the system or balking, that is, leaving the system without being served. It is assumed that a customer's decision to join or balk is irrevocable, and she is not allowed to renege. The joining/balking decision is made by evaluating a utility function that is commonly assumed to have the following form:

$$U = v - C\left(w\right) \tag{8.1}$$

where v denotes the customer service valuation (also referred to as the willingness to pay or the reservation price in the economic literature), w is a measure of delay in the system, and $C\left(\cdot\right)$ is a non-decreasing cost function converting the delay disutility experienced by the customer to monetary units.[2] It is common to assume that the cost function has a linear form,[3] that is, $C\left(w\right) = c_w w$, in which c_w is the unit delay cost rate that reflects customer delay sensitivity, and w is the expected delay in the system, that is, the sum of delays in the queue and in the service.[4] In addition, we assume that the customers have identical delay sensitivities.[5]

Now, we assume that the customers cannot observe the queue length but that the steady-state distribution of the delay is common knowledge, and each customer can accurately calculate the expected delay for a given arrival rate.[6] The expected delay in the system as a function of the arrival rate and service rate $w\left(\lambda, \mu\right)$ is increasing in the arrival rate, that is, $\dfrac{\partial w\left(\lambda, \mu\right)}{\partial \lambda} > 0$, and decreasing in the service rate, that is, $\dfrac{\partial w\left(\lambda, \mu\right)}{\partial \mu} < 0$, for $\lambda < \mu$, and $w\left(\lambda, \mu\right) = \infty$ for $\lambda \geq \mu$ (see, e.g., Asmussen 2008). This implies that as the arrival rate increases, the delay cost increases and, consequently, the utility decreases. For a given service rate, the utility function as defined by Equation (8.1) is implicitly dependent on the arrival rate through the delay cost term, that is, $U\left(\lambda \mid \mu\right) = v - C\left(w(\lambda|\mu)\right)$. This dependence plays a pivotal role in the analyses in this section.

In the queueing literature, customers are largely assumed to be independent and rational in their decisions. Each customer decides to join the system only if her utility is non-negative.[7] A customer's decision obviously influences other customer decisions through the (expected) delay they experience and the resulting delay cost. At an aggregate level, the interactions between customers' decisions result in an equilibrium pattern of behavior. Assuming that customers are heterogeneous in their service valuations, the equilibrium arrival rate is the solution to

$$\lambda = \Lambda \Pr\left(U\left(\lambda\right) \geq 0\right) = \Lambda \bar{F}\left(C\left(w\left(\lambda\right)\right)\right) \tag{8.2}$$

in which $\bar{F}\left(x\right) = 1 - F\left(x\right)$ and $F\left(\cdot\right)$ is the cumulative distribution function of the customer service valuation. As one may notice, there is not necessarily a closed-form expression for the arrival rate (demand). The solution to Equation (8.2) provides the service provider with the customer choice behavior as a function of the problem's primitive data and allows him to maximize his profitability through different marketing or operational decisions.

When queue length is observable and the service time distribution is common knowledge, a customer arriving at a system with i customers already in the system can calculate her expected delay in the system as $\dfrac{i+1}{\mu}$. In this case, according to Larsen (1998), the equilibrium arrival rate can be found as $\lambda = \left(1 - q_0\left(\Lambda\right)\right)\mu$, in which $q_0\left(\lambda\right) = \left(\displaystyle\sum_{i=0}^{\infty} A_i\left(\lambda\right)\right)^{-1}$ is the probability that the system is empty, and

$q_i = A_i q_0$, $A_i = \dfrac{\Lambda}{\mu}\overline{F}\left(\dfrac{c_w i}{\mu}\right)A_{i-1}$ for $i \geq 1$, and $A_0 = 1$. The literature on the observable queues is rather limited compared to that of the unobservable queues; we therefore focus on the latter studies in the major part of this chapter.

8.3 FACTORS INFLUENCING CHOICE BEHAVIOR

8.3.1 FACTORS DIRECTLY INFLUENCING UTILITY

8.3.1.1 Price

Using price as a tool to control the congestion level dates back to the seminal work by Naor (1969). He seems to be the first to consider customers as decision-makers in congested systems. He observed that the self-selecting customers ignored the negative externalities[8] they imposed on others and overloaded the system, which led to deviations from what a social planner would prescribe. Naor (1969) proposed pricing as a tool to regulate customer behavior and reach a socially optimal welfare. Subsequent studies have used somewhat similar modeling approaches and have assumed that the customers are both price and delay sensitive. The following linear utility function has been commonly used to capture customers' sensitivity to price.

$$U = v - p - C\left(w\right) \tag{8.3}$$

where p is the *posted* price (see, e.g., Edelson and Hilderbrand 1975; Mendelson and Whang 1990; Besbes and Maglaras 2009; Afeche and Pavlin 2016, for a similar approach). According to Equation (8.3), a customer's utility from a service is the difference between her service valuation and the *full price* (sum of the posted price and delay cost) that she incurs to be served. Under the assumptions described in Section 8.2 and with the customer choice behavior modeled as in Equation (8.3), the revenue-maximizing service provider's optimization problem can be formulated as

$$\max_{p}\left\{R\left(p\right) = p\lambda\left(p\right): \overline{F}^{-1}\left(\frac{\lambda\left(p\right)}{\Lambda}\right) - p - C\left(w\left(\lambda\left(p\right)\right)\right) = 0\right\}$$

8.3.1.2 Speed

As mentioned in Section 8.2, the expected delay is decreasing in μ. This implies that increasing the service speed (service rate) can have a positive impact on customer choice behavior through decreasing the expected delay. However, the service speed can also influence the quality that is delivered to customers, particularly in situations where providing a service or product calls for a high level of assiduity. Hence, increasing the speed can also have a negative impact on customer choice behavior, as the quality may decline.

In many businesses, termed as *customer-intensive services*, striking a balance between speed and quality is crucial. Examples can be found in various service systems (health care services, consultancy services, high-end restaurants, beauty care, etc.) and manufacturing environments (see, e.g., Kostami and Rajagopalan 2014 for an example of a car accessory producer whose value proposition centers on designing and producing customized products to fit a particular vehicle with a short waiting time). In the queueing literature with strategic customers, Anand, Paç, and Veeraraghavan (2011) seem to be the first to model such an interaction through the following utility function

$$U = V(\mu) - p - C(w) \tag{8.4}$$

(see Dai, Akan, and Tayur 2017 for a similar approach used for outpatient services). $V(\mu)$ is a function that encapsulates the interaction between the service valuation and service speed. Anand, Paç, and Veeraraghavan (2011) use the following functional form for $V(\mu)$,

$$V(\mu) = \left(v - \alpha(\mu - \mu_0)\right)^+ \tag{8.5}$$

where $x^+ = \max(x, 0)$, $\alpha \geq 0$ captures the customer intensity of the service provided, and μ_0 is the benchmark service speed. Obviously, a higher value of α means that customers react more intensely to deviations from the benchmark speed and, therefore, it implies a stronger dependence of the service value on the service speed. The special case of $\alpha = 0$ restores the model to its classic form, where the service value does not depend on the service speed.

There are other functional forms in the literature that are used to model dependence of the service value on the service speed. For instance, Tong and Rajagopalan (2014) consider customers who are heterogeneous in their intensity and propose

$$V(\mu) = \sqrt{\frac{1}{\alpha}\left(\frac{1}{\mu} - \frac{1}{\mu_0}\right)^+} \tag{8.6}$$

where α is a random variable with a known and bounded distribution. Tong (2012) assumes that the customer valuation increases in the service time consumed, until they feel saturated, for example, in consumer services such as golf courses and ski rides. Accordingly, he proposes the following model to account for this observation:

$$V(\mu) = \frac{1}{\mu} - \frac{1}{\alpha \mu^n}, \quad \text{for } \alpha > 0 \text{ and } n > 1, \tag{8.7}$$

where the value of n represents how fast a customer feels saturated.

There are also empirical results supporting the existence of a relationship between service duration and quality. Kc and Terwiesch (2009) show that increasing the service speed can result in an increase in the likelihood of mortality in surgeries. Similarly, Kc and Terwiesch (2012) report an increase in the probability of future re-admissions when reducing the length of stay in an intensive care unit. Other empirical studies, for example, Gross et al. (1998), Lin et al. (2001), and Chen, Farwell, and Jha (2009), also demonstrate a link between higher customer satisfaction and an increase in service duration in health care services.

8.3.1.3 Reliability

While the queueing literature predominantly focuses on delivery time as the only delivery performance measure of the service provider, empirical evidence supports the fact that other aspects of the delivery performance, like delivery reliability, can also influence customer choice behavior. Delivery reliability is defined as the rate of on-time deliveries, that is, deliveries before or on the promised delivery time. Based on the data from a retailer's online shop, findings in Rao, Rabinovich, and Raju (2014) demonstrate a negative relationship between reliability of deliveries and the rate of returns. Also, observations in Rao, Griffis, and Goldsby (2011) show that a failure to fulfill the order within the promised time, results in a decrease in order quantity and order value, as well as an increase in anxiety, in future purchases.

There is a tight link between delivery time and delivery reliability and ceteris paribus, a tighter delivery time commitment would be achievable at the price of a lower delivery reliability. In other terms, although a shorter delivery time can be favorable for the customers, it can worsen the delivery performance in the reliability dimension and can therefore cause them disutility. Customer sensitivity to different dimensions of delivery performance, that is, delivery time and delivery reliability, makes the trade-off between these dimensions a focal point in several industries. Examples can be found in online retailing (Rao, Griffis, and Goldsby 2011), furniture manufacturing (Shang and Liu 2011), banking (Ho and Zheng 2004), and container shipping (Jalili Marand, Li, and Thorstenson 2020). This trade-off can be modeled using the following utility function:

$$U = V(r) - p - C(w) \tag{8.8}$$

in which r denotes the delivery reliability, and $V(r)$ captures the dependence of the service valuation on the delivery reliability. One common approach is to model the service valuation as a linear function of the delivery reliability as

$$V(r) = v + \beta r \tag{8.9}$$

where $\beta \geq 0$ reflects the customer's sensitivity to the delivery reliability level (see, e.g., Ho and Zheng 2004; Jalili Marand, Li, and Thorstenson 2020; Shang and Liu 2011, for a similar approach). Jalili Marand, Tang, and Li (2019) consider the following specific functional form:

$$V(r) = v + \beta \left(\frac{1}{r_0} - \frac{1}{r} \right) \tag{8.10}$$

in which r_0 is the benchmark reliability in the industry, and β measures customer sensitivity to deviations from this benchmark. The benchmark values in different industries are reported by, for example, third-party industrial analysis providers. For instance, the benchmarks for electronics, personal care, and pharmaceutical products have been reported to be 91%, 88.7%, and 90.2%, respectively (Shang and Liu 2011).

8.3.1.4 Location
In some applications, customers need to travel to get the desired service. Traveling brings about customer disutility both because of the cost they have to incur for traveling and because of the time they need to spend commuting. This is the case in applications where the customers are geographically dispersed, but the service facilities are/need to be located at fixed locations. The sensitivity of customers to distance becomes important when the service provider has to decide on facility locations, knowing that location decisions will influence the load on the system through customer choice behavior. Such cases are common in health care systems (see, e.g., Dan and Marcotte 2019; Zhang et al. 2010 for examples of walk-in clinic and mammography network design problems, respectively).

In this case, customer choice behavior can be modeled using the following utility function:

$$U = v - p - C(w) - G(d) \tag{8.11}$$

in which d is a measure of the distance that a customer needs to travel to get the desired service, and $G(\cdot)$ is a non-decreasing cost function converting the distance disutility to monetary units (see, e.g., Andritsos and Aflaki 2015; Dan and Marcotte 2019; Dobson and Stavrulaki 2007; Etebari 2019; Hassin, Nowik, and Shaki 2018; Tong 2011 for a similar approach). The simplest form of the distance cost function is a linear form as

$$G(d) = g + \gamma d \tag{8.12}$$

where γ denotes customer distance sensitivity (see, e.g., Pangburn and Stavrulaki 2008 for an example). Note that distance may measure the actual spatial separation between the customer and the service provider, but it can also be used to reflect the difference between a customer's ideal service and that provided by the service provider (see Hotelling 1990 for more details).

8.3.2 DEPARTURE FROM THE RATIONALITY ASSUMPTION

Naor (1969) and the subsequent studies in the queueing literature, which consider customers to be decision-makers, predominantly assume they are rational; they are perfect utility-maximizers forming accurate expectations regarding the costs and benefits of different alternatives. However, there is abundant empirical evidence supporting the fact that customers are either limited in their computational abilities or prone to some unobserved noisy bias (Davis 2018), such that they may end up with non-optimal decisions, even though they attempt to maximize their utilities. Such behaviors are often referred to as *bounded rationality* in the economic literature (see, e.g., Simon 1957). Departing from the *full rationality* assumption has been one of the recent trends emerging in the queueing literature. In this section, we review some of the attempts made in modeling boundedly rational customers' choice behavior in congested systems. Interested readers are referred to Ren and Huang (2018) for a review of different approaches for modeling customer bounded rationality in operations management.

8.3.2.1 Logit Choice Models

One of the factors influencing customer choice behavior is customers' cognitive ability to calculate the expected utilities from different alternatives they are faced with and to compare them. Luce (1959) shows that as a result of inconsistency in comparative judgments, a decision-maker (here, a customer) who is given a set of alternatives from which to choose is more likely to choose the better alternatives over the worse ones. This probabilistic behavior can be modeled using a logit choice model. Considering the interactive behaviors of boundedly rational decision-makers in a game theoretic setting, McKelvey and Palfrey (1995) conceptualize the *quantal response equilibrium* assuming that each decision-maker follows the logit choice model and believes others also do so.

Huang, Allon, and Bassamboo (2013) seem to be the first to use such an approach to explicitly model boundedly rational customer choice behavior in a congested system. They consider a customer who needs to choose between two alternatives: buying the service from a service provider or balking. Due to the inconsistency in comparative judgments, her perceived utility from each alternative may differ from her actual utility. The difference, which reflects a bias in customer choice behavior, can emanate from the customers' inability to accurately estimate the expected delay in the system and consequently to estimate their expected utility from buying the service. This difference can be modeled using a random variable as follows:

$$\mathcal{U} = U + \epsilon \tag{8.13}$$

in which $\mathcal{U}$ and U are the perceived (expected) utility and the actual (expected) utility from buying the service,[9] and ϵ is a random variable with logistic distribution with mean zero and variance $\frac{\eta^2 \pi^2}{3}, \eta > 0$. U can follow one of the forms introduced in Section 8.3.1. At an aggregate level, the effective arrival rate to the system can be found from the equation

$$\lambda = \Lambda \left(\frac{e^{\frac{U}{\eta}}}{e^{\frac{U}{\eta}} + 1} \right) \tag{8.14}$$

in which U is implicitly dependent on λ through the delay cost function. Note that the scale parameter of the logistic distribution, η, is interpreted as a measure of customer bounded rationality level in this case. As η increases, the customers become more boundedly rational, and as it approaches zero, the customers' choice behavior approaches the rational behavior. Alternatively, the random variable can be used to model heterogeneity of preferences among a population of customers, uncertainty in choice outcomes due to unobservable variables affecting a given customer's choice, or variety-seeking and deliberately altering choice behavior of customers (see Talluri and Van Ryzin 2004, Chapter 7, for more details).

Using a linear price- and delay-dependent utility function, similar to Equation (8.3), within a logit choice model, Huang, Allon, and Bassamboo (2013) study social welfare and revenue maximizing pricing decisions of a monopolistic service provider. They show that ignoring the customer bounded rationality can lead to substantial revenue and welfare losses, even when the bounded rationality level is low. The logit choice model has also been used with other functional forms of the utility. Examples can be found in Li, Guo, and Lian (2016) and Li et al. (2017) that use Equation (8.4), Jalili Marand, Li, and Thorstenson (2020) that use Equation (8.8), and Etebari (2019) and Dan and Marcotte (2019) that employ Equation (8.11).

The *multinomial logit model* is an extension of the logit model and is used when there are more than two alternatives from which customers can choose, for instance, when there are multiple service providers in the market providing substitute services and the customers can choose to buy the service from one of the service providers or balk without being served. In such cases, the interactions between service providers can be modeled as a game (see, e.g., Allon and Federgruen 2007; Ho and Zheng 2004; Shang and Liu 2011 for a few examples), and each service provider's demand is determined by

$$\lambda_i = \Lambda \left(\frac{e^{\frac{U_i}{\eta}}}{\sum_{j=1}^{N} e^{\frac{U_j}{\eta}} + 1} \right) \tag{8.15}$$

in which λ_i is the arrival rate at service provider i, $i \in 1,\ldots,N$, N is the number of service providers competing in the market, and U_i is the customer utility from the service provided by service provider i.[10] In a competitive setting, So (2000) studies the interactions between an arbitrary number of service providers employing a log-linear utility function

$$U_i = \ln v_i p_i^{-c_p} w_i^{-c_w} \tag{8.16}$$

in which v_i, p_i, and w_i denote service valuation, price, and delay, respectively, at service provider i, and c_p is a measure of customer price sensitivity. Using the utility function in Equation (8.16) within the logit modeling framework results in the following aggregate demand model:

$$\lambda_i = \Lambda \left(\frac{v_i^{\eta} \, p_i^{\frac{1}{\eta} \, -c_p} \, w_i^{\frac{-c_w}{\eta}}}{\sum_{j=1}^{N} v_j^{\eta} \, p_j^{\frac{1}{\eta} \, -c_p} \, w_j^{\frac{-c_w}{\eta}} + 1} \right) \tag{8.17}$$

which is known as the multiplicative competitive interaction model in the literature (see, e.g., Huang, Leng, and Parlar 2013 for more details).

8.3.2.2 Anecdotal Reasoning

Other customers' experiences can also influence a given customer's perception of service quality. In congested systems, the calculation of the expected utility may be difficult for a customer due to inherent uncertainties in the system or limited available information. For instance, when a customer's interactions with a system (e.g., a restaurant or a clinic) are very infrequent, her knowledge of the system specifications (e.g., the service quality) would be scarce. In such cases, the customer may rely on anecdotes from other customers to form her expectations about the service quality and to make her buying/balking decision. Social media provide an easily accessible platform for exchanging such experiences.[11] The anecdotal reasoning framework, proposed by Osborne and Rubinstein (1998), can be used to capture decision-makers' anecdotal reasoning. According to this framework, a customer samples a service by asking prior users and then decides on her possible participation based on the sample average.

Using the anecdotal reasoning framework, Ren, Huang, and Arifoglu (2018) study a case in which customers are uncertain about service quality (reflected in the service valuation). In their study, a customer gathers k samples upon her arrival, and these samples are assumed to be drawn independently from the service valuation distribution with mean v and standard deviation σ. Each customer's estimation of service valuation can therefore be calculated as $\hat{v} = \dfrac{\sum_{i=1}^{k} v_i}{k}$. When the service valuation distribution is normally distributed, $\hat{v} \sim N\left(v, \dfrac{\sigma^2}{k}\right)$. The parameter k measures the customer bounded rationality level: As the number of anecdotes gathered for estimation (k), increases, the sample average deviates from the actual mean to a lower degree, since the sample variance decreases in k. In this case, the effective arrival rate to the system can be found by solving the following equation:

$$\lambda = \Lambda \Pr(U \geq 0) = \Lambda \bar{\Phi}\left(\frac{-\sqrt{k}U}{\sigma} \right) \tag{8.18}$$

in which $\bar{\Phi}$ is the complementary cumulative distribution function of the standard normal distribution. Note that U is implicitly dependent on λ through the waiting cost function. A similar approach is used by Huang and Chen (2015) in a situation where customers use anecdotal reasoning when they do not know the service rate and thus lack the capability to accurately estimate the expected delay.

8.3.2.3 Loss Aversion and Reference Dependence

Empirical evidence supports the fact that a customer's perceived utility from a service is mainly a result of her expectations (see, e.g., de Oña and de Oña 2015; Kuo and Jou 2014; Lin, Lee, and Jen 2008). These expectations form *reference points* based on which the customer compares the service in question and makes her buying/balking decision. Such a behavior, referred to as a *reference-dependent* behavior, can be studied through the lens of prospect theory (Kahneman and Tversky 1979). According to prospect theory, a decision-maker, who finds herself in a risky situation, thinks of her utility in terms of gains and losses in comparison to the reference points rather than to absolute values, and she perceives losses more intensely than equal-sized gains (known as *loss aversion*).

Yang, Guo, and Wang (2018) seem to be the first to study loss averse reference-dependent customer choice behavior in a congested system. They assume that a price- and time-sensitive customer's utility is the sum of their intrinsic utility and gain-loss utility where the first part is similar to the classic queueing literature (e.g., Edelson and Hilderbrand 1975) and the second part accounts for the additional utility she derives from comparing the actual outcome with the reference points. They let q_e denote the equilibrium joining probability of the customer. They define the reference points for the net monetary reward and for the delay as random variables $\hat{v}$ and $\hat{w}$, respectively, where $\hat{v} = v - p$ with probability q_e and $\hat{v} = 0$ with probability $1 - q_e$, and $\hat{w} = W$ with probability q_e and $\hat{w} = 0$ with probability $1 - q_e$. W is a random variable representing the delay in the system. Accordingly, they calculate the customers' joining and balking utilities and define $g(q_e)$ as the difference between the utilities of joining and balking when other customers join with probability q_e as

$$g(q_e) = U_{join} - U_{balk} = (2 - q_e + \alpha q_e)(v - p) - c_w w \left(1 + \beta - \frac{q_e(\beta - 1)}{2}\right) \quad (8.19)$$

where $\alpha \geq 1$ and $\beta \geq 1$ measure the degree of customer loss aversion related to net monetary reward and delay, respectively, and w is the expected delay in the system. They show that if $g(0) \leq 0$, then the balking with probability 1 is the pure strategy, if $g(1) \geq 0$, then the joining with probability 1 is the pure strategy; otherwise customers adopt a mixed strategy where $q_e \in (0,1)$ solves $g(q_e) = 0$.[12] In the latter case, the equilibrium arrival rate can be found as a function of the service provider's pricing decision. In a double-sided matching queue, Jiang et al. (2020) use a similar approach to model loss aversion in customer choice behavior.

8.3.3 ATTITUDE TO RISK

Customers' attitudes toward risk is another factor influencing their preferences and thus their purchase decisions. The classic queueing literature predominantly assumes risk neutral customers. However, some studies deviate from this assumption. In queueing systems, customers' risk aversion (risk seeking) is captured through structural assumptions regarding the delay cost function and/or utility function. From the delay cost perspective, Van Mieghem (2000) is the first to consider the delay cost as a general convex function increasing in delay. The convexity of this function indicates the customers' risk aversion with respect to delay. This assumption is also in line with the theory of the psychological cost of waiting. Osuna (1985) posits that after waiting for a certain amount of time, a decision-maker's anxiety and stress increase. The underlying reason is the feeling of waste and uncertainty about the remaining waiting time. Thus, the marginal psychological cost of waiting for an individual increases with the waiting time. As an example of an explicit nonlinear cost function, Kumar and Randhawa (2010) assume the cost function with the form of $C(w) = hw^r$ for some $r > 0$ and $h \geq 1$. Here, the parameter r specifies the cost function curve; for $r > 1$ it will be a convex function, whereas $0 < r < 1$ yields a concave function.

Although the customers in the mentioned studies are not risk neutral in terms of *delay uncertainty*, their utility is still risk neutral in relation to the resulted *delay cost*. The first study in which customers are assumed to be risk averse with respect to the net benefit of service dates back to Chen and Frank (2004). In this study, a general utility function is considered as follows:

$$U = f\left(v - p - C(w)\right) \tag{8.20}$$

in which $f(\cdot)$ is strictly increasing and concave. The authors assume homogeneous customers, that is, customers use the same utility function and their valuation of service, charged price, and cost function are identical. Extending Equation (8.20), Benioudakis, Burnetas, and Ioannou (2021) include delay compensation in the utility function. The main idea is that the service provider quotes a lead time d, and whenever the total delay exceeds d, the service provider compensates the customer by paying back l monetary value per unit of overrun. Therefore, we have the following utility function:

$$U = f\left(v - p - C(\tilde{w}) + l(\tilde{w} - d)^+\right) \tag{8.21}$$

in which $\tilde{w}$ is the realized delay. Afèche, Baron, and Kerner (2013) use the following functional form for the utility:

$$U = f\left(v - P(\tilde{w}) - C(\tilde{w})\right) = 1 - e^{-\alpha\left(v - P(\tilde{w}) - C(\tilde{w})\right)} \tag{8.22}$$

in which $\alpha > 0$ is the customer's risk tolerance, C is a linear function, and $P(\tilde{w}) = \alpha - \beta\tilde{w}$, with $\alpha, \beta \geq 0$, is the price function that depends on the realized delay. The pricing scheme decreases the risk of achieving a negative realized payoff for customers since the service provider compensates for the delays by charging lower prices. The authors assume heterogeneous customers with different service valuations but identical utility and cost functions.

While previous studies assume customers to be risk averse with respect to the total net benefit of being served, Wang and Zhang (2018) consider both risk-averse and risk-seeking customers and compare the effects of their risk attitudes on the joining strategies. The authors use a mean-variance utility function, which is well established in finance. This function is represented as $U(X) = E[X] - \frac{1}{2}AVar(X)$, where the value of A describes the subject's risk attitude such that $A > 0$, $A = 0$, and $A < 0$ indicate risk aversion, risk neutrality, and risk seeking, respectively. The utility function considered in Wang and Zhang (2018) turns out to be as follows:

$$U = v - c_w w - \frac{1}{2} A c_w^2 Var(W) \tag{8.23}$$

in which w and $Var(W)$ are the expected delay and the variance of delay, respectively.

8.3.4 INFORMATION

One of the important factors influencing customer choice behavior is the level of information available to customers at the time of making the joining/balking decision. A customer arriving at a restaurant, for instance, may see the number of seated and waiting customers but not know the actual service times or the quality of the service, or a caller may be informed about the expected delay by the call center but have no access to the number of callers ahead of her in the line. A service provider can use information availability to control the queue and improve the system performance. Even though the impact of information on customer choice behavior and system performance has been addressed in the queueing literature, there is no unified well-established terminology in place. In this section, we address this subject from two perspectives: delay information and quality information.[13]

8.3.4.1 Delay Information

Guo and Zipkin (2007) consider three levels of delay information that a service provider can decide to reveal to customers who are heterogeneous in their delay sensitivities. In particular, they consider the following three cases: no information,[14] partial information,[15] and full information.[16] Assuming that customers are not sensitive to price, they set the system throughput and average customer utility as the performance measures of the service provider and the customers, respectively. They explore the conditions under which revealing accurate delay information can bilaterally benefit the customers and the service provider and find that the form of the delay cost rate distribution plays a crucial role. They particularly show that revealing more information always results in throughput improvement, given that customers are heterogeneous enough in their delay sensitivities. Among others, Guo and Zipkin (2008, 2009a, 2009b), Economou and Kanta (2008), and

Hu, Li, and Wang (2018) have extended Guo and Zipkin (2007) in various directions. Hu, Li, and Wang (2018), for instance, focus on the interaction between fully informed and uninformed customers and the resulting system performance and show that a larger proportion of informed customers may not necessarily contribute to throughput or social welfare.

While the mentioned studies all assume that customers are aware of the service rate or a distributional form of it, Cui and Veeraraghavan (2016) consider a *blind queue* in which customers have heterogeneous beliefs about the unknown service rate or have been systematically misinformed. They use a distribution-free approach and study the impact of service-rate information revelation on the performance of a queueing system. They conclude that the impact of the service-rate information on congestion and welfare is mixed. Particularly, they show that although a service provider's revenue improves upon announcing its service rate under certain conditions, the congestion level also increases, which implies an increased market coverage at the cost of worsened customer welfare. For a health care service, for instance, the implication could be that the revelation of service-rate information can enhance service accessibility and simultaneously diminish the utility of the visiting patients through the increase in the waiting time. Note that despite the impact that the delay-information revelation can have on the performance of service systems, most of the queueing literature focuses on the no-information case, according to Guo and Zipkin's (2007) terminology.

8.3.4.2 Quality Information

There are situations where the service provider cannot or has no incentive to communicate the service quality to customers in a dependable way. In such situations, customers who are not informed about the quality would infer it from other sources, for example, the number of customers waiting in the queue or from anecdotes from earlier customers (see anecdotal reasoning in Section 8.3.2.2). For instance, long lead times for an innovative product or a long waiting line in a restaurant can signal service/product quality for an uninformed customer. In some of the pioneering works, Debo, Rajan, and Veeraraghavan (2012) and Debo, Parlour, and Rajan (2012) propose modeling frameworks of inferring unknown quality from congestion and price in congested systems, respectively. In a laboratory experiment, Kremer and Debo (2016) verify the qualitative results of Debo, Rajan, and Veeraraghavan (2012) and show that under certain conditions, uninformed customers' purchase probability can locally increase in the waiting time. They also provide experimental evidence explaining the *empty restaurant syndrome*, where uninformed customers, despite lower delay costs, tend to avoid shorter waiting times (an empty restaurant) as they infer a lower quality from such a service (food quality). Other extensions can be found in, for example, Ren, Huang, and Arifoglu (2018), He, Chen, and Righter (2020), and Guo et al. (2020). Although the impact of quality information on the choice behavior of heterogeneous customers has recently drawn more attention in academic works, most of the classic queueing literature has mainly assumed that customers are informed about the service quality.

8.4 CONCLUSIONS AND FUTURE RESEARCH OPPORTUNITIES

This chapter provides an overview of a range of approaches used to model customer choice behavior in congested systems (see Table 8.1 for a summary of the common functional forms of utility/demand surveyed in this chapter). We have aimed to stress the fact that the performance of a congested system at an aggregate level

TABLE 8.1

Summary of the Common Functional Forms of Utility/Demand Used in the Queueing Literature

Factor		Utility/demand function	Examples
Factors directly influencing the utility	Price	$U = v - p - C(w)$	Naor (1969), Edelson and Hilderbrand (1975)
		$U = \ln vp^{-c_p} w^{-c_w}$	So (2000)
	Speed	$U = \left(v - \alpha(\mu - \mu_0)\right)^+ - p - C(w)$	Anand, Paç, and Veeraraghavan (2011), Li, Guo, and Lian (2016), Liu et al. (2018)
		$U = 1 - \alpha e^{-\frac{1}{\mu}} - p - C(w)$	Tong, Nagarajan, and Cheng (2016)
		$U = \sqrt{\frac{1}{\alpha}\left(\frac{1}{\mu} - \frac{1}{\mu_0}\right)^+} - p - C(w)$	Tong and Rajagopalan (2014), Tong, Nagarajan, and Cheng (2016)
		$U = \frac{1}{\mu} - \frac{1}{\alpha\mu^n} - p - C(w)$	Tong (2012)
	Reliability	$U = v + \beta r - p - C(w)$	Ho and Zheng (2004), Jalili Marand, Li, and Thorstenson (2020), Shang and Liu (2011)
		$U = v + \beta\left(\frac{1}{r_0} - \frac{1}{r}\right) - p - C(w)$	Jalili Marand, Tang, and Li (2019)
	Location	$U = v - p - C(w) - g - \gamma d$	Pangburn and Stavrulaki (2008), Hassin, Nowik, and Shaki (2018), Andritsos and Aflaki (2015), Dan and Marcotte (2019)
Bounded rationality	Logit	$\lambda = \Lambda\left(\frac{e^{\frac{U}{\eta}}}{e^{\frac{U}{\eta}} + 1}\right)$	Huang, Allon, and Bassamboo (2013), Li, Guo, and Lian (2016), So (2000)

(Continued)

TABLE 8.1

Summary of the Common Functional Forms of Utility/Demand Used in the Queueing Literature

Factor		Utility/demand function	Examples
	Anecdotal reasoning	$\lambda = \Lambda \bar{\Phi} \left(\dfrac{-\sqrt{k}U}{\sigma} \right)$	Ren, Huang, and Arifoglu (2018)
	Loss aversion	λ solving $\left(2 - \dfrac{\lambda}{\Lambda} + \alpha \dfrac{\lambda}{\Lambda} \right)(v-p) -$ $c_w \left(\dfrac{1}{\mu - \lambda} \right)\left(1 + \beta - \dfrac{(\beta-1)\lambda}{2\Lambda} \right) = 0$	Yang, Guo, and Wang (2018)
Risk aversion		$U = v - p - c_w w^r$	Kumar and Randhawa (2010)
		$U = 1 - e^{-\alpha\left(v - P(\hat{w}) - C(\hat{w}) \right)}$	Afèche, Baron, and Kerner (2013)
		$U = v - c_w w - \dfrac{1}{2} A c_w^2 Var(w)$	Wang and Zhang (2018)

largely depends on how delay, as the main service/product attribute, is perceived by the system users both in isolation and in interaction with other service/product attributes. It is safe to conclude that a better understanding of customer choice behavior in congested systems cannot be achieved without making use of contributions to the field of the psychology of waiting. This also explains why the queueing literature, as reviewed in this chapter, mainly uses models of individual customer choice rather than models of aggregate demand.

The interplay between theoretical studies and empirical research creates enormous research opportunities. On the one hand, given that the choice models covered in this chapter are widely applied in a range of studies using various assumptions regarding the customer and/or system-related parameters from a theoretical point of view (see Hassin 2016; Hassin and Haviv 2003 for two excellent reviews on different theoretical models developed in the queueing literature), there is obviously a need for intensified empirical endeavors to test the extent to which these purely theoretical results—often using hypothetical assumptions—would actually be applicable in practice. Moreover, we see an abundance of theoretical works within the queueing literature that have generalized results on customer behavior empirically shown to hold at the individual level and transferred these results to the aggregate behavior at the system level without sufficient justification.

On the other hand, the rapidly growing body of literature on the psychology of waiting by the increased availability of observational field and experimental data (see, e.g., Allon and Kremer 2018 for an excellent review of recent developments) calls for queue theoretical models that incorporate the advancements in the behavioral fields into the formal operations management literature. Such models would shed light on how different factors influencing customer choice behavior in congested systems are applicable as levers by operations managers to improve the system performance.

NOTES

[1] For unobservable queues, it is sufficient to assume that the service discipline is strong and work conserving.

[2] In other contexts, the delay cost can be considered a surrogate for the loss of goodwill or can be associated with the actual cost of holding jobs in the system (e.g., work-in-process holding cost, particularly if the jobs require a large amount of space or specialized condition) or to pertain to the opportunity cost associated with waiting customers who cannot generate revenue (see Allon, Bassamboo, and Gurvich 2011 for examples).

[3] The functional form of the cost can reflect the customers' attitude toward risk, as we discuss in Section 8.3.3.

[4] Alternatively, it can be considered to be the queueing time or delivery time quoted by the service provider.

[5] This is not an essential assumption. In fact, the literature on customer heterogeneity is rich. But they use similar utility functions to those we introduce in this chapter to capture different customer segment choice behaviors.

[6] In Section 8.3.2, we address departures from the rationality assumption.

[7] In general, they will join if their utility is greater than or equal to a certain threshold value. Here, without loss of generality, the threshold value is normalized to zero.

[8] The delay that an arriving customer who joins the queue imposes on customers arriving later.

[9] The perceived utility from balking is modeled in a similar way, assuming that the actual utility from balking is normalized to zero.

[10] Note that the random noise term is Gumbel distributed in the multinomial logit model, and η is the scale parameter of the corresponding distribution.

[11] Note that such anecdotes may not be an accurate representation of the situation that a customer would actually encounter.

[12] Note that this model makes a strong connection between the reference points used to calculate the utilities for buying or balking and the strategy of other customers. As a result, a customer who deviates from the common behavior risks a greater distance from the reference point and incurs a larger loss. Consequently, the queue is polarized such that long queues become even longer and short queues even shorter (a.k.a. herding behavior).

[13] The interested reader is referred to Hassin (2016), Chapter 3, for a comprehensive review.

[14] Similar to the study by Edelson and Hilderbrand (1975) in which customers only estimate the distribution of the waiting time and thereby their expected delay in the system based on long-term equilibrium behavior.

[15] Similar to the study by Naor (1969) in which the customers know the number in the queue but are uncertain about the actual service times of the waiting customers.

[16] In this case, the customers know the exact waiting times.

REFERENCES

Afèche, Philipp, Opher Baron, and Yoav Kerner. 2013. Pricing time-sensitive services based on realized performance. *Manufacturing & Service Operations Management* 15, no. 3: 492–506.

Afeche, Philipp, and J. Michael Pavlin. 2016. Optimal price/lead time menus for queues with customer choice: Segmentation, pooling, and strategic delay. *Management Science* 62, no. 8: 2412–2436.

Allon, Gad, Achal Bassamboo, and Itai Gurvich. 2011. "We will be right with you": Managing customer expectations with vague promises and cheap talk. *Operations Research* 59, no. 6: 1382–1394.

Allon, Gad, and Awi Federgruen. 2007. Competition in service industries. *Operations Research* 55, no. 1: 37–55.

Allon, Gad, and Mirko Kremer. 2018. Behavioral foundations of queueing systems. *The Handbook of Behavioral Operations* 9325: 323–366.

Anand, Krishnan S., M. Fazıl Paç, and Senthil Veeraraghavan. 2011. Quality—speed conundrum: Trade-offs in customer-intensive services. *Management Science* 57, no. 1: 40–56.

Andritsos, Dimitrios A., and Sam Aflaki. 2015. Competition and the operational performance of hospitals: The role of hospital objectives. *Production and Operations Management* 24, no. 11: 1812–1832.

Asmussen, Søren. 2008. *Applied Probability and Queues*. Vol. 51. New York: Springer Science & Business Media.

Benioudakis, Myron, Apostolos Burnetas, and George Ioannou. 2021. Lead time quotations in unobservable make-to-order systems with strategic customers: Risk aversion, load control and profit maximization. *European Journal of Operational Research* 289, no. 1: 165–176.

Besbes, Omar, and Costis Maglaras. 2009. Revenue optimization for a make-to-order queue in an uncertain market environment. *Operations Research* 57, no. 6: 1438–1450.

Chen, Hong, and Murray Frank. 2004. Monopoly pricing when customers queue. *IIE Transactions* 36, no. 6: 569–581.

Chen, Lena M., Wildon R. Farwell, and Ashish K. Jha. 2009. Primary care visit duration and quality: Does good care take longer? *Archives of Internal Medicine* 169, no. 20: 1866–1872.

Cui, Shiliang, and Senthil Veeraraghavan. 2016. Blind queues: The impact of consumer beliefs on revenues and congestion. *Management Science* 62, no. 12: 3656–3672.

Dai, Tinglong, Mustafa Akan, and Sridhar Tayur. 2017. Imaging room and beyond: The underlying economics behind physicians' test-ordering behavior in outpatient services. *Manufacturing & Service Operations Management* 19, no. 1: 99–113.

Dan, Teodora, and Patrice Marcotte. 2019. Competitive facility location with selfish users and queues. *Operations Research* 67, no. 2: 479–497.

Davis, Andrew M. 2018. Biases in individual decision-making. *The Handbook of Behavioral Operations*: 149–198.

Debo, Laurens G., Christine Parlour, and Uday Rajan. 2012. Signaling quality via queues. *Management Science* 58, no. 5: 876–891.

Debo, Laurens G., Uday Rajan, and Senthil K. Veeraraghavan. 2012. Signaling by price in a congested environment. *Ann Arbor* 1001: 48109.

de Oña, Juan, and Rocio de Oña. 2015. Quality of service in public transport based on customer satisfaction surveys: A review and assessment of methodological approaches. *Transportation Science* 49, no. 3: 605–622.

Dobson, Gregory, and Euthemia Stavrulaki. 2007. Simultaneous price, location, and capacity decisions on a line of time-sensitive customers. *Naval Research Logistics (NRL)* 54, no. 1: 1–10.

Economou, Antonis, and Spyridoula Kanta. 2008. Optimal balking strategies and pricing for the single server Markovian queue with compartmented waiting space. *Queueing Systems* 59, no. 3–4: 237.

Edelson, Noel M., and David K. Hilderbrand. 1975. Congestion tolls for Poisson queuing processes. *Econometrica: Journal of the Econometric Society*: 81–92.

Etebari, Farhad. 2019. A column generation algorithm for the choice-based congested location-pricing problem. *Computers & Industrial Engineering* 130: 687–698.

Gross, David A., Stephen J. Zyzanski, Elaine A. Borawski, Randall D. Cebul, and Kurt C. Stange. 1998. Patient satisfaction with time spent with their physician. *Journal of Family Practice* 47, no. 2: 133–138.

Guo, Pengfei, Moshe Haviv, Zhenwei Luo, and Yulan Wang. 2020. Optimal queue length information disclosure when service quality is uncertain: A Bayesian persuasion approach. SSRN 3599693.

Guo, Pengfei, and Paul Zipkin. 2007. Analysis and comparison of queues with different levels of delay information. *Management Science* 53, no. 6: 962–970.

Guo, Pengfei, and Paul Zipkin. 2008. The effects of information on a queue with balking and phase-type service times. *Naval Research Logistics (NRL)* 55, no. 5: 406–411.

Guo, Pengfei, and Paul Zipkin. 2009a. The effects of the availability of waiting-time information on a balking queue. *European Journal of Operational Research* 198, no. 1: 199–209.

Guo, Pengfei, and Paul Zipkin. 2009b. The impacts of customers' delay-risk sensitivities on a queue with balking. *Probability in the Engineering and Informational Sciences* 23, no. 3: 409–432.

Hassin, Refael. 2016. *Rational Queueing*. Boca Rotan: Chapman and Hall/CRC.

Hassin, Refael, and Moshe Haviv. 2003. *To Queue or Not to Queue: Equilibrium Behavior in Queueing Systems*. Vol. 59. Norwell, MA: Kluwer Academic Publishers.

Hassin, Refael, Irit Nowik, and Yair Y. Shaki. 2018. On the price of anarchy in a single-server queue with heterogeneous service valuations induced by travel costs. *European Journal of Operational Research* 265, no. 2: 580–588.

He, Qiao-Chu, Ying-Ju Chen, and Rhonda Righter. 2020. Learning with projection effects in service operations systems. *Production and Operations Management* 29, no. 1: 90–100.

Ho, Teck H., and Yu-Sheng Zheng. 2004. Setting customer expectation in service delivery: An integrated marketing-operations perspective. *Management Science* 50, no. 4: 479–488.

Hotelling, Harold. 1990. Stability in competition. In *The Collected Economics Articles of Harold Hotelling*, 50–63. New York: Springer-Verlag.

Hu, Ming, Yang Li, and Jianfu Wang. 2018. Efficient ignorance: Information heterogeneity in a queue. *Management Science* 64, no. 6: 2650–2671.

Huang, Jian, Mingming Leng, and Mahmut Parlar. 2013. Demand functions in decision modeling: A comprehensive survey and research directions. *Decision Sciences* 44, no. 3: 557–609.

Huang, Tingliang, Gad Allon, and Achal Bassamboo. 2013. Bounded rationality in service systems. *Manufacturing & Service Operations Management* 15, no. 2: 263–279.

Huang, Tingliang, and Ying-Ju Chen. 2015. Service systems with experience-based anecdotal reasoning customers. *Production and Operations Management* 24, no. 5: 778–790.

Jalili Marand, Ata, Hongyan Li, and Anders Thorstenson. 2020. Competing on price, speed, and reliability: How does bounded rationality matter? *Journal of the Operational Research Society*: 1–14.

Jalili Marand, Ata, Ou Tang, and Hongyan Li. 2019. Quandary of service logistics: Fast or reliable? *European Journal of Operational Research* 275, no. 3: 983–996.

Jiang, Tao, Xudong Chai, Lu Liu, Jun Lv, and Sherif I. Ammar. 2020. Optimal pricing and service capacity management for a matching queue problem with loss-averse customers. *Optimization*: 1–24.

Kahneman, Daniel, and Amos Tversky. 1979. Prospect theory: An analysis of decision making under risk. *Econometrica* 47, no. 2: 263–291.

Kc, Diwas S., and Christian Terwiesch. 2009. Impact of workload on service time and patient safety: An econometric analysis of hospital operations. *Management Science* 55, no. 9: 1486–1498.

Kc, Diwas Singh, and Christian Terwiesch. 2012. An econometric analysis of patient flows in the cardiac intensive care unit. *Manufacturing & Service Operations Management* 14, no. 1: 50–65.

Kostami, Vasiliki, and Sampath Rajagopalan. 2014. Speed—quality trade-offs in a dynamic model. *Manufacturing & Service Operations Management* 16, no. 1: 104–118.

Kremer, Mirko, and Laurens Debo. 2016. Inferring quality from wait time. *Management Science* 62, no. 10: 3023–3038.

Kumar, Sunil, and Ramandeep S. Randhawa. 2010. Exploiting market size in service systems. *Manufacturing & Service Operations Management* 12, no. 3: 511–526.

Kuo, Chung-Wei, and Rong-Chang Jou. 2014. Asymmetric response model for evaluating airline service quality: An empirical study in cross-strait direct flights. *Transportation Research Part A: Policy and Practice* 62: 63–70.

Larsen, Christian. 1998. Investigating sensitivity and the impact of information on pricing decisions in an M/M/1/∞ queueing model. *International Journal of Production Economics* 56: 365–377.

Li, Xin, Pengfei Guo, and Zhaotong Lian. 2016. Quality-speed competition in customer-intensive services with boundedly rational customers. *Production and Operations Management* 25, no. 11: 1885–1901.

Li, Xin, Qingying Li, Pengfei Guo, and Zhaotong Lian. 2017. On the uniqueness and stability of equilibrium in quality-speed competition with boundedly-rational customers: The case with general reward function and multiple servers. *International Journal of Production Economics* 193: 726–736.

Lin, Chen-Tan, Gail A. Albertson, Lisa M. Schilling, Elizabeth M. Cyran, Susan N. Anderson, Lindsay Ware, and Robert J. Anderson. 2001. Is patients' perception of time spent with the physician a determinant of ambulatory patient satisfaction? *Archives of Internal Medicine* 161, no. 11: 1437–1442.

Lin, Jiun-Hung, Tzong-Ru Lee, and William Jen. 2008. Assessing asymmetric response effect of behavioral intention to service quality in an integrated psychological decision-making process model of intercity bus passengers: A case of Taiwan. *Transportation* 35, no. 1: 129–144.

Liu, Yixuan, Xiaofang Wang, Stephen Gilbert, and Guoming Lai. 2018. Pricing, quality and competition at on-demand healthcare service platforms. SSRN 3253855.

Luce, R. Duncan. 1959. *Individual Choice Behavior: A Theoretical Analysis*. New York: Wiley.

McKelvey, Richard D., and Thomas R. Palfrey. 1995. Quantal response equilibria for normal form games. *Games and Economic Behavior* 10, no. 1: 6–38.

Mendelson, Haim, and Seungjin Whang. 1990. Optimal incentive-compatible priority pricing for the M/M/1 queue. *Operations research* 38, no. 5: 870–883.

Naor, Pinhas. 1969. The regulation of queue size by levying tolls. *Econometrica: Journal of the Econometric Society*: 15–24.

Osborne, Martin J., and Ariel Rubinstein. 1998. Games with procedurally rational players. *American Economic Review*: 834–847.

Osuna, Edgar Elias. 1985. The psychological cost of waiting. *Journal of Mathematical Psychology* 29, no. 1: 82–105.

Pangburn, Michael S., and Euthemia Stavrulaki. 2008. Capacity and price setting for dispersed, time-sensitive customer segments. *European Journal of Operational Research* 184, no. 3: 1100–1121.

Rao, Shashank, Stanley E. Griffis, and Thomas J. Goldsby. 2011. Failure to deliver? Linking online order fulfillment glitches with future purchase behavior. *Journal of Operations Management* 29, no. 7–8: 692–703.

Rao, Shashank, Elliot Rabinovich, and Dheeraj Raju. 2014. The role of physical distribution services as determinants of product returns in Internet retailing. *Journal of Operations Management* 32, no. 6: 295–312.

Ren, Hang, and Tingliang Huang. 2018. Modeling customer bounded rationality in operations management: A review and research opportunities. *Computers & Operations Research* 91: 48–58.

Ren, Hang, Tingliang Huang, and Kenan Arifoglu. 2018. Managing service systems with unknown quality and customer anecdotal reasoning. *Production and Operations Management* 27, no. 6: 1038–1051.

Shang, Weixin, and Liming Liu. 2011. Promised delivery time and capacity games in time-based competition. *Management Science* 57, no. 3: 599–610.

Simon, Herbert Alexander. 1957. *Models of Man: Social and Rational; Mathematical Essays on Rational Human Behavior in Society Setting*. New York: Wiley.

So, Kut C. 2000. Price and time competition for service delivery. *Manufacturing & Service Operations Management* 2, no. 4: 392–409.

Stidham, Shaler. 2009. *Optimal Design of Queueing Systems*. Boca Rotan: Chapman and Hall/CRC.

Talluri, Kalyan T., and Garrett J. Van Ryzin. 2004. *The Theory and Practice of Revenue Management*. Vol. 68. USA: Springer.

Tong, Chunyang. 2012. Pricing schemes for congestion-prone service facilities. *Operations Research Letters* 40, no. 6: 498–502.

Tong, Chunyang, Mahesh Nagarajan, and Yuan Cheng. 2016. Operational impact of service innovations in multi-step service systems. *Production and Operations Management* 25, no. 5: 833–848.

Tong, Chunyang, and Sampath Rajagopalan. 2014. Pricing and operational performance in discretionary services. *Production and Operations Management* 23, no. 4: 689–703.

Tong, Dehui. 2011. Optimal pricing and capacity planning in operations management. PhD Thesis, University of Toronto.

Van Mieghem, Jan A. 2000. Price and service discrimination in queuing systems: Incentive compatibility of Gc μ scheduling. *Management Science* 46, no. 9: 1249–1267.

Wang, Jinting, and Zhe George Zhang. 2018. Strategic joining in an M/M/1 queue with risk-sensitive customers. *Journal of the Operational Research Society* 69, no. 8: 1197–1214.

Yang, Liu, Pengfei Guo, and Yulan Wang. 2018. Service pricing with loss-averse customers. *Operations Research* 66, no. 3: 761–777.

Zhang, Yue, Oded Berman, Patrice Marcotte, and Vedat Verter. 2010. A bilevel model for preventive healthcare facility network design with congestion. *IIE Transactions* 42, no. 12: 865–880.

9 Dynamic Innovation Capabilities

U.K. uz Zaman

Department of Mechatronics Engineering, College of Electrical and Mechanical Engineering, National University of Sciences and Technology, Islamabad, Pakistan

A. Naseem

Department of Engineering Management, College of Electrical and Mechanical Engineering, National University of Sciences and Technology, Islamabad, Pakistan

CONTENTS

9.1 INTRODUCTION: BACKGROUND AND DRIVING FORCES

Dynamic capabilities are distinct from ordinary capabilities because they focus on strategic management. Therefore, firms can further strengthen ordinary capabilities by aggregating them with dynamic capabilities, which will maintain and extend their competitive advantage. The ordinary capabilities of a firm enable it to perform efficiently, but when dynamic capabilities are combined with an effective strategy, the firm can produce the right products to meet customer demands and target the right markets for competitive and technological opportunities in the future. Dynamic capabilities support the firm in developing speculations to validate and then realign resources/assets as required. Consequently, strong dynamic capabilities are crucial for the success and growth of a firm, especially when it positions itself in a high-tech market.

Moreover, innovation is considered a main factor that triggers not only productivity and economic growth but also acts as a major source of employment (Fagerberg, Mowery, and Nelson 2005). The innovation paradigm (Yin, Ming, and Zhang 2020) has evolved over the years. It started with the closed innovation paradigm (Innovation 1.0) and moved to the collaborative or symbiosis innovation paradigm (Innovation 2.0), to the open innovation paradigm (Innovation 3.0), and finally to the co-innovation paradigm (Innovation 4.0). The innovations at the level of firms have been found to create new jobs in advanced as well as developing countries (Harrison et al. 2014). Such innovations not only promote sustainable growth, they also improve productivity. Therefore, companies/firms/industries need to consider competence and knowledge simultaneously to gain a competitive advantage. One way can be through the launching of new products (Laursen and Salter 2006). Such quick launch of new products not only secures new turnover, it also provides appropriate business avenues for competitors and helps firms gain new market shares (O'Cass and Ngo 2011; Teece 2007). Therefore, the ability of a firm to introduce/generate new products is crucial for all stakeholders.

Keeping in view the mentioned factors, the firm's dynamic capabilities can be exploited with different types of capabilities in a dynamic environment, and it is more significant for the firms in high-tech industries. The Fourth Industrial Revolution (termed as Industry 4.0) is integrating current manufacturing technologies with new generation information technologies (Haag and Anderl 2018). Global initiatives such as Made in China 2025, Industries 4.0 in Germany, and Advanced Manufacturing Partnership in the United States, are using intelligent manufacturing to gain competitive advantages for the manufacturing industry of major countries (Zhou et al. 2018; Zhong et al. 2017).

Consequently, the use of technology has changed not only the products but also its processes and services. This can affect other firms directly or indirectly, based on the integration of firms with other firms. It is mandatory for such firms to focus on innovation and modify the behaviors to continue to prosper. Consequently, this chapter focuses on three wide areas governing dynamic innovation capabilities: innovation in agile businesses, innovation in sensing and seizing capabilities, and applications of dynamic capabilities with perspective of manufacturing. Each area is introduced in this chapter along with the relevant state-of-the-art of dynamic innovation capabilities and supported by case studies.

9.2 INNOVATION IN AGILE BUSINESS SYSTEMS

9.2.1 Demand as an Opportunity

Demand is a harvesting activity and demand management is a vivacious process in which information about new needs, ideas, and projects are collected internally and externally. The internal information assists in making strategies aligned with portfolios. For demand management, organizations should adopt a proactive approach. It should be the responsibility of top management. The demand-management process can be effective and useful in identifying the real strategic commitment of the organization. It helps the organization in developing the correct portfolio. The strategy provides the directions and selection models that describe the best set of tasks aligned with strategic objectives and effectively meet the demand requirements (Gildea and Foster 2018).

Rapid advancements in information technologies, including Internet of Things, data analytics, and artificial intelligence, have transformed traditional manufacturing into an intelligent one (Rialti et al. 2019). Intelligent manufacturing focuses on self-optimization and autonomy. It manages demand effectively by recommending models based on learning and cognitive capacities. It supports dynamic knowledge-based skills and autonomous manufacturing by intelligent (1) perceiving, (2) simulating, (3) understanding, (4) predicting, (5) optimizing, and (6) controlling strategies. Moreover, it supports the capacities of (1) self-thinking, (2) self-decision-making, (3) self-executing, and (4) self-improving (Zhou et al. 2019).

9.2.2 Making the Transition via Agile

In the IT world, it is necessary to think about Agile. If any organization is not working on it, it is behind the curve. Demand management generally follows a waterfall approach in which leaders are responsible for such planning. It helps in the identification and allocation of resources based on demands. Demand management, a critical facet of Agile, is a core feature of management. Agile management is comprised of processes for making plans and managing the current and forecasted demands (Gildea and Foster 2018). The Agile approach shifts the focus from individual projects funding to full scope of the products and value for the organization by fulfilling requirements and delivering to customers. Agile is a framework in which there are defined processes and methodologies.

The model for successful business is built on the effective management of demand. The Agile methodologies allow organizations to manage demand and accelerate deployment and development cycles because of Agile's ability and flexibility. Therefore, businesses all over the world are embracing Agile. In Agile implementations, it is difficult for quality-assurance managers to handle user stories and requests because such requirements can complicate matters due to their effects on lines of business and numerous geographies.

9.2.3 Agile Attributes

In the real world, demand keeps changing every single day, which is a challenge for organizations. When demand management is effective, managers can respond to such dynamic changes efficiently. Agile assists managers with capturing the demand as early as possible and in effective ways. It helps in ensuring the necessary amendments to meet the demands at the right time, subsequently saving resources. Hence, Agile, in the demand-management process, leads to appropriate continual planning, forecasting, and resourcing. The main attributes of Agile methods are user orientation, iterative procedure, and high flexibility. Strong user orientation is one of the special features of the Agile method. It is accomplished with the provision of a functional prototype and validations of the requirements from customers at systematic intervals. In this way, any amendment of deviation can be assimilated in the development process without any delay. It also lessens the risk of developing such products that are not acceptable by the customer. Other attributes of Agile methods, including high flexibility and an iterative approach, are appropriate for the developments in prompt advancements (Shams et al. 2020). The attributes of agile methods are shown in Figure 9.1.

9.2.4 Exploitation, Exploration, and Agile

Despite high pressure for innovation, it is difficult for organizations to establish strategic innovation management to meet customer demand requirements because there are limited resources in every organization. Innovation is an activity parallel to day-to-day business. Generally, when there is a good economic situation in organizations, it reinforces the commitment of resources and more concentration in day-to-day business other than planned innovation tasks. The challenge exists mostly in the instantaneous management of the incremental and disruptive innovations. Here, there is a need for exploration and exploitation—exploring new possibilities and exploiting existing possibilities (Niewohner et al. 2019; Breznik, Lahovnik, and Dimovski 2019). The exploitation focuses on productivity, efficiency, and stability, while exploration targets flexibility and growth (Keller and Weibler 2015), as shown in Figure 9.2.

Particularly, exploration and exploitation include searching and finding new means and methods and rethinking existing approaches. In addition, exploration also

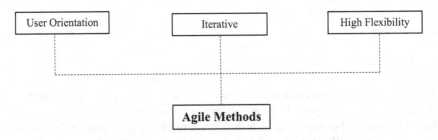

FIGURE 9.1 Attributes of Agile methods (adopted from Niewohner et al. 2019).

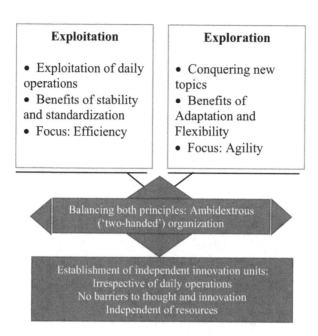

FIGURE 9.2 Balance between exploitation and exploration (Weibler/Keller 2015).

includes processes agility. Both strategies have different risk factors. For example, in exploitation, issues in safety, stability, and standardization are considered the main risk factors, while exploration assumes to opt for higher risks via flexibility and adaptability. The Agile model supports the development of radical innovations (Keller and Weibler 2015). In such a scenario, organizations work on establishing independent innovation units along with their daily operations. There are no barriers to new innovations, and resources will also be independent (Niewohner et al. 2019).

9.2.5 STRATEGIC AGILITY IN AN INTERNATIONAL BUSINESS CONTEXT

When an organization wants to be agile in an international business context, there are few key operational areas on which to focus. The agility is not considered a stand-alone proficiency; rather, it results from a mass of competences that integrate speed, adaptability, sustainability, innovation, and organizational resilience. Hence, for an organization to become an agile multinational organization, the integration of agility in various key operational areas is required. These are (1) information technology, (2) supply chain, and (3) sustainable production. Although these are different streams, they complement each other and help organizations grow into more agile forms (Shams et al. 2020), as shown in Figure 9.3.

Strategic agility operates with IT capabilities to enhance agility. In supply chains, IT capabilities are required within the complete chain. Therefore, it can be stated that IT boosts the collective capabilities that are positively and indirectly related with strategic agility. It makes IT vital for achieving the strategic agility in a complete

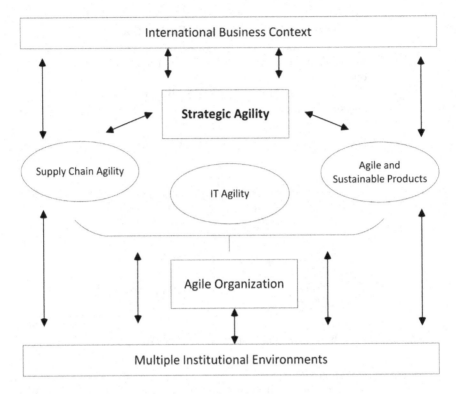

FIGURE 9.3   Framework for agile multinational firms (Shams et al. 2020).

supply chain. It is similar in the case of supply chain versions. In this context, there are three key strategic factors associated with agility, including IT agility, sustainable production, and supply chain agility (Alsaad, Yousif, and AlJedaiah 2018). For an organization to achieve agility and success, these factors play essential roles. IT agility is basically the superior IT skills and capabilities that can assist in improving organizational performance. If IT is widely used in multinational organizations, it becomes easier for managers to communicate across geographical and functional boundaries. It improves coordination and enhances performance by exploiting market threats and exploring opportunities. IT competence augments organizational agility. It allows for operational adjustment agility and market capitalizing agility (Ravichandran 2018). For instance, big data analytics capabilities improve the decision-making process (Rialti et al. 2019). When it is combined with knowledge management, performance is allied with it (Ferraris et al. 2019) because IT competence can create a positive impact on organizational performance by flexibility and reactive processes and operations (Ravichandran 2018; Shams et al. 2020).

Supply chain agility is one of the fundamental drivers of strategy in multinational firms. The supply chain promptly responds to market demand and customer requirements, aiming to achieve agile operations because it is a competence that allows the organization to meet the unpredictable and constantly changing conditions. Hence, it affects the cost efficiency and customer effectiveness. The fundamental enablers of supply chain agility are: quality of supplier relationship, a high level of the shared information, and a high level of connectivity among the organizations in a supply chain (Alsaad, Yousif, and AlJedaiah 2018). In the relationship between integration and information, the important factor of customer or marketing sensitivity has a dynamic capability (Shams et al. 2020).

Sustainable production is significant in coupling with agility and agile productions. In the modern era, multinational firms are asked to convert their production process to sustainable operations because of market vitality and growing product complexity. Although it seems that sustainability and agility are two different terms, their integration in manufacturing and/or production leads organizations to superior reactivity to external stimuli and an improved position in the national and/or international market. Hence, agility is the main performance measure of the agile manufacturing process. It creates an organization's ability to not only have variability of the products but also act in a cost-effective manner within a short span of time. However, sustainability focuses on the minimization of business impacts on the environment. Therefore, the integration of these two factors results in several benefits, including time compression, cost effectiveness, value-added product variety, and existence in a competitive business environment (Shams et al. 2020).

An example is explained in the following text about the benefits of including agile attributes in businesses. A firm with 350 employees and 50 years of history was not responding to quick advancements to avail opportunities in the industry. The firm felt that they were not providing value to their customers. Therefore, they might have been losing members as well. In such a scenario, a team was developed to identify the key patterns in the firm's approach to innovation. They developed definite amendments in the technology and processes to help the firm respond to changes more quickly and provide value to their customers. The team found that the firm was not concentrating on innovation and agility. the team focused on two building blocks of agility: One was to eliminate activities that were not contributing to the company's goals and its ability to embrace amendments. Traditionally, change is managed very carefully to avoid any critical disruptions (Lasach 2019), but the team agreed that the firm did not need such caution.

In innovation, the team observed that there was no dearth of creativity and inspiration in the firm. But there was a lack of taking initiative to change the processes and approaches. The firm was working with traditional approaches of development and testing. While the firm was proud of their 50 years of history, the team recognized that the firm needs to focus on changes right now, particularly with respect to the industry in which they were operating. The firm needed to strategize, change its current approach, and stop activities that were not adding value (Helfat and Peteraf 2009; Stoyanova 2018). So, the firm implemented software in which any employee from any department could suggest change and innovation within the company. Moreover, the firm revised its project evaluation process by adding external perspectives. A few years after the implementation, it was identified that data-driven decision-making processes incorporating innovation and agility brought remarkable outcomes, including effective and well-timed customer demand management and cost-effectiveness.

9.3 INNOVATION IN SENSING, SEIZING, AND RECONFIGURING CAPABILITIES

9.3.1 CUSTOMER DEMAND AND SUPPLY CHAIN

A company's consumer value-based philosophy has been more popular than ever before. It implies that firms can only outperform rivals if they can build a strategic edge that they can retain. A competitive advantage can be achieved by delivering greater benefits to consumers by presenting the same advantages at a reduced cost (cost advantage) as rivals, and by providing advantages that outweigh those of competing offers. Such incentives are also known as competitive advantages because they decide the location of the organization in the industry. In market planning, the strategic advantage is formulated and made sustainable, which should be impossible for rivals to neutralize. First, the motive is the company's capacity to separate itself from competitors in terms of goods and supply chain capacities (or customer service) and, second, to work at lower prices with less personnel and thus more flexibility. It encourages the company to create superior profitability and better income for itself with its consumers and owners. Superior output (or productivity) is thus the outcome of cost-effectively producing and providing superior consumer value, which is achieved by organizing the business around understanding how cost-effectively consumer value is created (managing value creation), how cost-effectively customer value is distributed (managing value delivery). Therefore, organizations need not only address and organize these procedures within their own organization but also across corporate borders (Lasach 2019). In addition, companies need to consider not just what each consumer values but also the forces behind changing value expectations over time, as value changes not just across consumers but also over time. The perception that companies have both a value production method (or demand chain) and a product distribution system (or supply chain) to be controlled and organized to optimize productivity and effectiveness is well known in the literature. A demand chain can be demarcated as a network of autonomous entities engaged in processes of demand to respond by value development to customer demand (Helfat and Peteraf 2009). Inside each company, all the demand processes and practices involved in creating consumer demand are part of the demand chain. The processes may vary amongst different types of organizations, but they usually include business intelligence, branding, promotion and distribution, and new product development (NPD) (Teece 2012). Demand chain management (DCM) seeks to control and organize certain market processes to recognize, produce, and increase consumer demand as cost-effectively as possible, within a single business and through the demand chain. Although DCM is, to a degree, cost-oriented (efficiency), by defining, developing, promoting, and selling attractive (advanced, personalized, and affordable) goods, the focus is on sales (effectiveness). A sales level that helps the individual business and the demand chain to succeed in the long term is the greatest measure of DCM quality (Chong and Zhou 2014). An innovative supply chain can be characterized as a network of independent entities engaged in supply processes to satisfy consumer demand by providing value. The supply chain comprises all the supply systems and operations involved with consumer demand fulfillment within each company. The procedures can vary between different types of organizations but generally involve acquisition, production, and delivery. Supply chain management (SCM) is directed at controlling and organizing these supply chains within a single organization and across the supply chain to satisfy consumer demand as cost-effectively as possible. While SCM focuses on sales (effectiveness) to a certain extent, the emphasis is on lowering costs (efficiency) by reducing the overall amount of money needed to deliver the appropriate quality of customer support. A cost and service quality that helps the individual organization and the supply chain to succeed in the long term is the supreme indicator of SCM excellence. Regarding the network of companies concerned (besides the focal business, these networks may include manufacturers, transporters, distributors/retailers, and the consumers themselves), there is little substantial distinction between the demand and supply chain, however, with regard to the mechanisms considered. The market and supply chain can also be interpreted as separate viewpoints on the same network of companies in this context. The term production supply chain (or value chain) will be used where the demand and supply procedures are concurrently considered. The reach starts with the center of supply (or point of origin) and ends at the point of consumption, independent of the chain viewpoint. This so-called ultimate chain, however, is rarely regarded in reality; instead, more narrow views are typically used by businesses. To achieve superior consumer service as cost-effectively as possible, differentiated supply chain management (DSCM) attempts to integrate demand and supply processes within a single organization and throughout the supply chain of demand. The focus is both on growing sales (effectiveness) through the provision of attractive goods and customized supply chain strategies (customer service) and on reducing costs (efficiency) through cost-effective control of demand and supply processes. A differentiated supply chain (also called a multiple supply chain) calls for differentiated (or tailored) customer service (Hilletofth 2012). In comparison to a "one size fits all" supply chain, a diverse supply chain includes multiple innovative supply chain solutions (mixtures of various procurement, production, and delivery options), each tailored for a single commodity or business environment. DSCM's aim is to achieve a strategic edge (increase profitability) by offering a lower cost for greater consumer satisfaction (Hilletofth 2012).

9.3.2 DYNAMIC CAPABILITIES MODEL

Teece's dynamic capabilities model offers tactics and strategies to build firms' capabilities as dynamic capabilities (Teece 2007, 2009, 2012). It surrounds sensing, seizing, and reconfiguring the dynamic capabilities via management innovation where sensing is the identification and evaluation of an opportunity, seizing is basically mobilizing the resources to address that opportunity and capturing value from it, and reconfiguring is continued renewal process (see Figure 9.4). All these activities are necessities for the firm to sustain itself because of technological advancements and continuous change in customer demand (Lasach 2019). In this context, one of the most important managerial functions is asset composition that comprised of its alignment, coalignment, realignment, and redeployment. Asset composition can be performed on a continuous or periodic basis. It is necessary to avoid or decrease internal conflict and increase value for the enterprise.

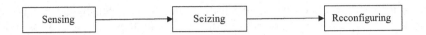

FIGURE 9.4 Teece's vision (Teece 2012).

Sensing capabilities recognize and deal with opportunities as well as threats. Seizing capabilities exploit the opportunities that were sensed and moreover fend off the threats (Breznik, Lahovnik, and Dimovski 2019; Teece 2012). Reconfiguring capabilities maintain competitiveness through enriching, combining, defending, and altering operational capabilities. Such types of capabilities introduce strategic change (Lasach 2019). Sensing the novel opportunities entails scanning and seeking processes associated with the business environment. Basically, sensing capability is considered market-focused learning. It states the firm's market-hunting efforts. It is allied with those processes that allow a firm to foresee market developments as well as customer requirements. Seizing encapsulates how companies tackle the sensed opportunities with organizing appropriate activities and defining the most appropriate business model to exploit the opportunities. It also signifies the benefits of investments that were realized in the sensed opportunities. Reconfiguring supports firms to constantly rearrange the operational capabilities with seized opportunities (Teece 2009, 2012). It is embedded with internally focused learning (Zollo and Winter 2002). The internal learning includes unlearning the existing operational capabilities and forming novel operational capabilities (Breznik, Lahovnik, and Dimovski 2019).

Nokia could not catch the revolution of the smartphone. The main reason is that the company was not well prepared and equipped for such sensing. Therefore, they missed the transformation of state-of-the-art technology. On the other hand, Apple not only sensed the customers' requirements but was also on the cutting edge of technological advancements. The company seized the opportunities and threats and built all the required capabilities step by step. It then transformed its tangible and intangible assets, renewed its core competencies, and developed new customer value propositions. For example, for work on the iPod, Apple developed capabilities in digital rights management and handheld device design. The company learned to cut deals with recording companies and studies. The team focused on user-friendly technology with attractive form factors. Consequently, Apple engaged with the reconfiguration of activities and resources to match customer requirements and demand of the changing environment.

Sensing and seizing capabilities are comparatively basic functions, while reconfiguring is more intricate and might require a business model to be entirely revamped (Teece 2007). The main idea of this breakdown of dynamic capabilities is to explain that how dynamic capabilities are deployed, developed, and manifested (Cirjevskis 2019; Bocken and Geradts 2020; Zollo and Winter 2002). In this context, dynamic capability is a "meta-capability" that surpasses ordinary firm capability. The difference between sensing, seizing, and reconfiguring is built on the idea that different management innovations have distinct attributes (Teece 2012). The attributes include frequency, causal ambiguity, and heterogeneity (Breznik, Lahovnik, and Dimovski 2019). The frequency explains the management routine, how often it is executed in a defined period. Causal ambiguity defines the relationship among actions and performance. Heterogeneity is associated with the range and types of activities essential for success with management innovations (Stoyanova 2018). The contribution of management innovations toward sensing, seizing, and reconfiguring depends on the varying degree of these three attributes. Sensing is considered by high heterogeneity, as it encompasses various kinds of tasks. Although management routines that drive the sense of opportunities and related threats are continuous, they have low causal ambiguity with respect to the action–performance nexus. The sensed opportunities and threats are seized occasionally, which indicates low frequency. The heterogeneity is measured in terms of the variety of activities involved in the management routines. It supports seizing; therefore, it remains comparatively high. There is higher causal ambiguity in seizing the sensed opportunities and threats in comparison with sensing. Reconfiguration can be done radically or continuously. Continuous reconfiguration suggests high frequency and high causal ambiguity among management routines and results and moderate heterogeneity of the execution tasks. A radical reconfiguration suggests low frequency, low causal ambiguity, and high heterogeneity.

For example, in a molding machine manufacturing company, to evaluate the service revenue and the profitability potential, a novel management routine was considered. The company introduced a variation to the cost accounting systems. It was identified that without tracking the profitability and service revenue, the company was never able to assign resources effectively for an extension of service business. Moreover, without testimony that the maintenance services were highly profitable, it was not possible to convince management of the need for the recruitment and training of more internal maintenance specialists. The manager stated that without changing parts of the company's accounting system and presenting the service activity–based costing scheme, it was not possible to extend the service offerings. This system would support the company in recognizing the financial advantages of services, track the costs and profitability of those services, and evaluate the financial returns from the company's investments. The composition of dynamic capability is explained in Table 9.1.

TABLE 9.1
Dynamic Capability Composed of Sensing, Seizing, and Reconfiguring Capabilities

The Composition of Dynamic Capability		
Sensing capability	**Seizing capability**	**Reconfiguring capability**
Companies need to explore the internal as well as external environments to find opportunities	When opportunities are sensed, they must be addressed via novel processes, products, and services	While addressing new opportunities, companies must recombine and reconfigure the capabilities and resources because of environmental changes
Common activities/practices: • Identification of new technologies • Identification of new ideas • Identification of new customers/markets	Common activities/practices: • Selection of the "right" new technology or business model • Building commitment and loyalty	Common activities/practices: • Stimulation for open innovation • Management of strategic fit • Deployment of knowledge management

Source: Breznik, Lahovnik, and Dimovski (2019).

9.3.3 BARRIERS AND DRIVERS TO SUSTAINABLE BUSINESS MODEL INNOVATION

Sustainable Business Model Innovation (SBMI) is considered vital in solving pressing societal issues and addressing customer demand. SBMI has the potential to address such challenges, especially demand requirements. To develop dynamic capabilities for business model innovation, it is necessary to give importance to a firm's organizational design. Dynamic capabilities are key to SBMI, but to make them effective, the organizational design is an essential factor. There are different dynamics of organizational design that impede or strengthen dynamic capabilities. A comprehensive understanding of drivers and barriers at the different levels of the organization are shown in Figure 9.5. These levels are institutional, strategic, and operational and influence the dynamic capabilities requirement for SBMI (Bocken and Geradts 2020).

At the institutional level, barriers are rules, norms, and beliefs that guide the organizational behavior that leads to strategic importance regarding the functional strategy, short-term profitability, and manipulation of current business operations. Such strategic barriers transform into operational barriers, which encompass (1) functional excellence, (2) standardized innovation processes and procedures, (3) fixed resource planning and allocation, (4) incentive systems, and (5) financial performance metrics. There are obstacles on each level that affect a corporation's ability to sense opportunities for SBMI, seize them, and transform their existing business models (Bocken and Geradts 2020).

Also, at the institutional level, there are drivers of balancing shareholder and stakeholder value, accepting ambiguity, and valuing business sustainability. These drivers help in lessening the negative impact of (1) focus on shareholder value, (2) uncertainty avoidance, and (3) short-termism, which are considered obstacles to SBMI. The institutional drivers inspire the strategic importance on (1) collaborative innovation, (2) a strategic focus on SBMI, and (3) patient investments. Consequently, these strategic drivers enable the sensing, seizing, and transforming of SBMI and offer a counterbalance to strategic obstacles. Subsequently, operational drivers also facilitate the practices for the execution of strategic actions favorable to dynamic capabilities for SBMI. Operational-level drivers include (1) people capability development, (2) enabling innovation structure (Stoyanova 2018), (3) ring-fenced resources for SBMI, (4) incentive schemes for sustainability, and (5) performance metrics for sustainability (Bocken and Geradts 2020) (refer to Figure 9.5).

9.3.4 CASE STUDY—SAMSUNG'S ACQUISITION OF HARMAN

In 2016, Samsung declared the acquisition of American automotive technology manufacturer Harman International Industries, marking Samsung's venture into the automotive industry. This acquisition included combining the companies' business models and dynamic capabilities, which would eventually assist in developing new customer value propositions. In this way, Samsung could deliver their customers the "ultimate professional experience." While exploring the reinvention of Samsung's business model, two triggers of dynamic capabilities were identified (Cirjevskis

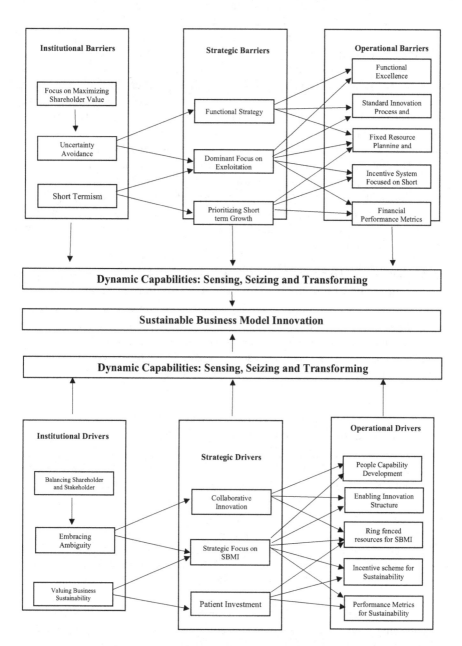

FIGURE 9.5 Barriers and drivers to sensing, seizing, and transforming SBMI (Bocken and Geradts 2020).

2019). The first one was Samsung's weak transformation capabilities because Samsung was not very successful in reconfiguring or transforming its resources. For example, after the loss of $5 billion because of the spontaneously combusting Note 7 device, Samsung tried something that would allow it to outshine its competitors and make the company a top smartphone manufacturer. Furthermore, Samsung, with the assistance of Harman's well-recognized market status, could plug into the field of automotive connectivity much faster than Apple or other rivals. It could bring actual innovation to this trailing, especially in the sector of car technology. The second trigger was the similarities and complementarities of Samsung's dynamic capabilities and the target companies. As Samsung's and Harman's technologies were communally complementary, they had a considerable market advantage. Similarities among the dynamic capabilities of Samsung and Harman include: successful sensing of the evolving market demands with respect to cars; seizing opportunities with the development of advanced products, services, and platforms; and keeping the top positions in their markets and getting competitive advantages. Therefore, Samsung's success in acquiring Harman was marked by the strong similarities between the two companies (Cirjevskis 2019).

9.3.4.1 Samsung's Dynamic Capabilities to Develop Electric Cars and/or Its Components before the Acquisition of Harman

Samsung sensed the opportunity that the automotive market was in the development phase of software-based cars. Samsung seized the opportunity and formed a team for its automotive electronic business in 2015 to find opportunities in a project titled 'connected business cars'. Then, Samsung invested in a Chinese company focused on electric car sales. Samsung created a team to develop the products for autonomous driving vehicles.

9.3.4.2 Harman's Dynamic Capabilities to Develop a Connected Car Solution before the Samsung Acquisition

Harman sensed that a connected car solution would be a great opportunity to gain operating profits by promoting additional features and services. Harman seized the opportunity by setting the standard in advancements in combination with an intuitive interface. Then, Harman transformed the resources and developed adaptive cruise control, collision avoidance, and warning systems.

Samsung's dynamic capabilities allowed them to reconfigure or transform the building blocks of its business model. Thus, Samsung sensed the new customer segments, that is, smart vehicles that extend the sophisticated embedded electronics and innovative key activities that should be built. Therefore, Samsung seized new vital resources (by acquiring Harman), a key partners' network, and Harman's customers. Consequently, Samsung reconfigured the market promotion channels and new customer segments. In this manner, Samsung scored the new customer value proposition with the provision of new subscriptions for their existing and new customers. So, it can be concluded that dynamic capabilities were the real drivers of innovation in Samsung's business model. These capabilities generated the new revenue stream, reduced cost, delivered a new value proposition, and created a new sustained competitive advantage (Cirjevskis 2019). Subsequently, bridging the

TABLE 9.2

Reinvention of Samsung Business Model and Micro-Foundations of Dynamic Capabilities

Reinvention of Samsung's business model	Micro-foundations of Samsung's dynamic capabilities
Sensing new customer segments	Samsung identified new customer demand and formed new key activities in the connected car manufacturing industry. Moreover, it sensed new initiatives required to satisfy this demand.
Seizing new resources	Samsung seized and assimilated valuable resources. They partnered with a firm that they acquired.
Transforming and reconfiguring new channels and customer relationships. The outcome is a new customer value proposition, cost structure, and revenue stream	Samsung generated a new revenue stream by transforming promotion channels to the connected car industry. The acquisition of Harman gave Samsung a strong position in the developing market, especially in automotive electronics.

Source: Adopted from Cirjevskis (2019).

perspectives of Samsung and Harman, Samsung's business model reinvented with the micro-foundations of dynamic capabilities, as shown in Table 9.2.

9.4 APPLICATIONS OF DYNAMIC INNOVATION CAPABILITIES WITH THE PERSPECTIVE OF MANUFACTURING

9.4.1 CHANGING THE MANUFACTURING PARADIGM

Manufacturing today is not just associated with the construction of physical products. Some of the most important decisions in the manufacturing industry are directly related to the nature of products, the economics of production, and fluctuations/changes in consumer demands. Around 70–80% of the cost incurred on a particular product is due to its manufacturability. It is imperative that a designer consider ease of manufacturing in the early design stages and provide a platform that is not only easy to follow in terms of manufacturing but also leads to reduced costs of logistics and assembly (Zaman et al. 2017). The concept of concurrent engineering (CE) stemmed from the same notion, which integrates actors that make decisions upstream to cater for requirements that are external and made downstream within the product-development process (Loch and Terwiesch 2000). In addition, design trade-offs related to product performance, such as productivity, utilization, producibility, and support, are also catered for by CE providing an "optimization" process.

With such dynamic environments, it is important to focus on the "complementarity" between product and process innovation (Guisado-González, Wright, and Guisado-Tato 2017) and further explore opportunities associated with it. Since both product and process innovation are now recognized to drive innovation activity equally, the benefits of considering them simultaneously can be many (Damanpour

2010). Hullova, Trott, and Simms (2016) proposed a conceptual model that explains the extents of product and process innovation in terms of complementarity (see Figure 9.6). The map can be used by team leads/managers to assist them in choosing the most suited complementarity strategy for a particular project.

Moreover, as the industries today are trying to implement Industry 4.0, intelligent manufacturing is taking place as the driver of dynamic innovation capabilities by introducing autonomy and self-optimization (Zhou et al. 2019). Such capabilities induce new demands, like cognitive capacities and learning for various industries and firms. In addition, smart manufacturing (Kusiak 2018) and cloud manufacturing (Adamson et al. 2017) have broadened the concept of intelligent manufacturing and have assisted in its implementation via methods such as RFID-based production monitoring (Guo et al. 2015), production data collection and analysis (Ding and Jiang 2018), and digital twin-based cyber–physical fusion (Ding et al. 2019).

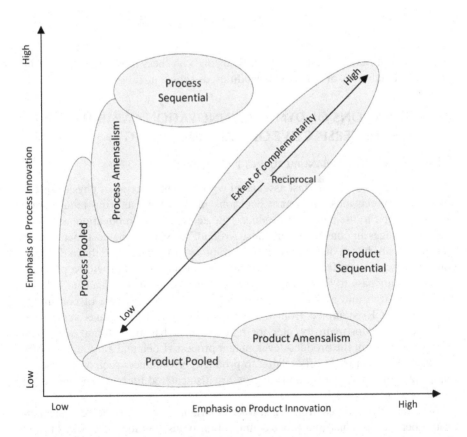

FIGURE 9.6 Product-process complementarity map to position a portfolio of projects. (Adopted from Hullova, Trott, and Simms 2016, p. 934.)

9.4.2 Sustainable and Smart Products

Products that are both smart as well as sustainable belong to a new generation of smart products that can help in achieving both sustainability and circular economy. Features such as connectivity, intelligence, sensing, human interaction, and autonomy are opening new doors for innovative capabilities within highly dynamic environments (Tomiyama et al. 2019). As possessing all of the mentioned capabilities is not easy, few firms can independently possess them in terms of innovation for sustainability. Multiple stakeholders need to come together to build an ecosystem around the products that can not only assist in gaining competitive advantages but also help in establishing partnerships with other firms for value co-innovation and co-creation (Yin, Ming, and Zhang 2020). Yin, Ming, and Zhang (2020) proposed five future directions to assist organizations with achieving high complementarity at the product-process level and making their systems as intelligent as possible: co-sharing associated with resource sharing challenges, co-construction for challenges of the emergence of co-innovation platform, co-creation with respect to outcome-related challenges, co-evolution for sustainability challenges, and co-existence for challenges of the stability (see Figure 9.7).

9.4.3 Case Study—Complementarity in Product-Process Innovation in the United Kingdom Food and Drink Industry (Hullova et al. 2019)

In every company's portfolio of projects targeting novel product and process developments, different complementarity types exist. The case study explained in this section focuses on dynamic innovation at the product-process level by studying how firms within the UK's food and drink industry should operate effectively to not only identify various types of complementarity between product and process innovation but also to manage new product and process development projects. Since complementarity type changes with every project in the portfolio, the project can be affected by various contingencies, such as more focus on product and process innovation, the project's level of innovativeness, and the existence of internal knowledge in product and/or process development (Hullova, Trott, and Simms 2016). Therefore, project teams need to develop a set of dynamic capabilities to help them in managing/controlling the contingencies in each project and to cater for such diverse kinds of complementarities. The cost of the project can be increased when time and resources are not allocated effectively toward a project, thereby leading to not only ineffective deployment of the workforce but also increased time to market (Bellgran 1998).

The authors used random selection to choose case studies from the food and drink industry that match four extents of complementarities between product and process innovation (process amensalism, product sequential, process pooled, and reciprocal). The findings of the case study were divided into three parts: (1) capacity to identify complementarity type, (2) ability to implement a suitable integration mechanism,

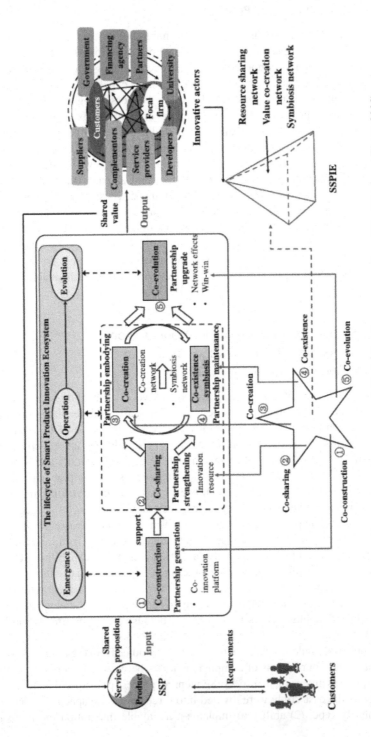

FIGURE 9.7 Life cycle of smart product innovation ecosystem with the five future perspectives (Yin, Ming, and Zhang 2020).

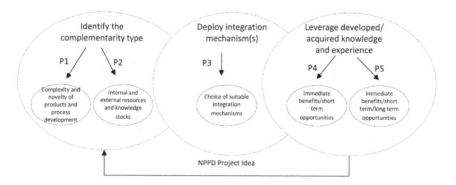

FIGURE 9.8 Framework to manage the complementarity between product and process innovation in new product process development (NPPD) projects (Hullova et al. 2019).

and (3) potential to utilize the developed/acquired knowledge and expertise in new product-process development projects. Furthermore, the authors concluded that the existing internal knowledge related to product and process development was critical to increasing the complementarity. However, as the complexity and novelty of the projects increased, there was a need for the identification of suitable external collaborators that could provide the knowledge that was missing internally. In light of all the findings, a framework was also proposed to manage the complementarity between product and process innovation in new projects, as shown in Figure 9.8. For instance, the projects that were termed as complex and involved a greater risk required higher integration, that is, reciprocal complementarity, whereas projects where the team personnel used their years of experience to increase the efficiency of the production lines could be categorized as low extent of complementarity (process-pooled) and are common in low-technology process industries, such as line stretch and bolt-ons.

9.4.4 CASE STUDY—KNOWLEDGE-DRIVEN DIGITAL TWIN MANUFACTURING CELL (ZHOU ET AL. 2019)

Manufacturing systems have evolved over the years, and their evolution can be divided into three stages based on degree of intelligence: traditional manufacturing, smart manufacturing, and intelligent manufacturing. Intelligent manufacturing has developed powerful learning and cognitive capabilities by integrating AI and advanced manufacturing with knowledge-driven decision-making (Zaman ct al. 2018) and knowledge-based intelligent process planning (Zhu et al. 2018). The case study in this section proposes a knowledge-driven digital twin manufacturing cell (KDTMC) that uses an intelligent simulating, perceiving, predicting, understanding, optimizing, and controlling strategy to maximize product throughput and quality and reduce cost by maintaining flexibility. The KDTMC is equipped with the

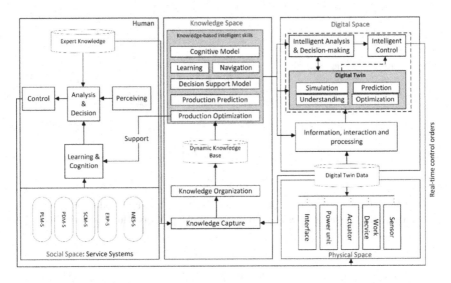

FIGURE 9.9   Architecture of KTDMC (Zhou et al. 2019).

capacities of self-thinking, self-decision making, self-execution, and self-improving. The architecture of KTDMC is shown in Figure 9.9.

The KTDMC, was applied to the intelligent manufacturing platform of Xi'an Jiaotong University, China, and displayed intelligent process planning, production scheduling, production process analysis, and dynamic regulation. It served as the minimum implementation unit for intelligent manufacturing.

9.5 LIST OF INNOVATION-DEPENDENT DEMAND FUNCTIONS

A list of innovation-dependent hybrid demand functions is listed in Table 9.3. It includes multiple parameters that constitute dynamic innovative capabilities, such as demand rate, demand magnitude, potential market size, initial market size, price per unit, innovation effect.

9.6 CONCEPTUAL FRAMEWORK FOR DYNAMIC
INNOVATION CAPABILITIES

Based on the overarching aim of the chapter and thorough understanding attained by studying the three main areas governing dynamic innovation capabilities—innovation in agile businesses, innovation in sensing and seizing capabilities, and applications of dynamic capabilities from the perspective of manufacturing—a conceptual framework is proposed, as shown in Figure 9.10. The framework contains two portions: enterprise and the external environment. To attain performance outcomes, it is imperative that dynamic capabilities such as sensing, seizing, and reconfiguring

TABLE 9.3

List of Innovation-Dependent Demand Functions

Demand function	Parameters, variables, and conditions	Reference
$\delta_m\left(z_t\right) = \eta_m z_t^{\varepsilon m}$	m: Product $\delta_m(z_t)$: Demand rate z_t: Customers are insensitive to the level of innovation η_m: Demand magnitude ξm: Growth rate	(Jana, Graves, and Grunow 2018)
$\lambda_t = p\left[\bar{N} - N(t)\right]$	$\bar{N}$: Potential market size $N(t)$: Cumulative number of adopters p: Coefficient of innovation λ_t: demand at time t	(Kumar, Alok, and Udayan Chanda 2016)
$f(t) = p(t)\left[1 - F(t)\right]$; where $F(t) = \dfrac{N(t)}{\bar{N}}$	$\bar{N}$: Potential market size $N(t)$: Cumulative number of adopters at time t $p(t)$: Innovation effect at time t	(Chanda and Kumar 2011
$\dfrac{d}{dt}U(t) = -\alpha U(t) - \beta\dfrac{U(t)A(t)}{M}$ $\dfrac{d}{dt}A(t) = \alpha U(t) + \beta\dfrac{U(t)A(t)}{M} - \pi A(t)$ $\bar{N}(t) = P(t)A(t)$ $\lambda(t) = \dfrac{dN(t)}{dt} = k\big(PA(t) - N(t)\big)$	M: Initial market size $U(t)$: Uninformed population at time t $A(t)$: Aware and favorable population at time t $\bar{N}$: Potential market size at time t $N(t)$: Cumulative number of adopters at time t P: Price per unit k: Initial adoption rate α: Innovation coefficient β: Imitation coefficient π: Fraction of the aware and unsatisfied population who left the market	(Chanda and Kumar 2019)

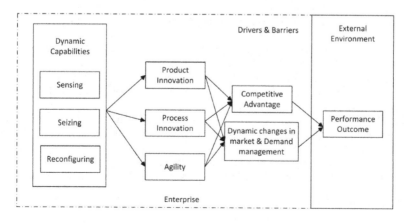

FIGURE 9.10 Proposed conceptual framework.

are combined with manufacturing innovation (at both the product level and process level) to gain a competitive advantage.

9.7 CONCLUSION AND FUTURE WORK

Dynamic capabilities enable firms to produce the right products and fulfill customer demand and target the right markets. When integrated with innovation, they promote sustainable growth, improve productivity, encourage the launch of new products, facilitate new turnover, provide employment opportunities, and invite an influx of superior knowledge and competence within the firms. The chapter provided a thorough understanding of the three main areas governing dynamic innovation capabilities: innovation in agile businesses, innovation in sensing and seizing capabilities, and applications of dynamic capabilities from the perspective of manufacturing. Each area was further supported by relevant case studies. By reviewing the relevant state-of-the-art of dynamic innovation capabilities, it can be deduced that if the policies and actions within firms are knowledge-driven, more sustainable business models can be created and implemented that will result in smart choices and decisions. Such outcomes will not only make decision-making more intelligent with respect to the selection of products and services, demand fluctuations, and changing product mix but will also assist in better employment opportunities.

As part of future research, various demand functions can be studied that incorporate innovation, customer awareness, and cost and inventory parameters. Furthermore, research can be conducted on demand functions, including the effect of external factors, particularly competitors' analysis and cultural effects, while adopting the innovated products considering technology transfer, adoption, and diffusion. In addition, studies can focus on the concept of digital twins, which not only complement product and process innovation but will also control dynamically changing demand parameters.

REFERENCES

Adamson, G., W. Lihui, H. Magnus, and P. Moore. 2017. Cloud manufacturing—a critical review of recent development and future trends. *International Journal of Computer Integrated Manufacturing* 30, no. 4–5: 347–380.

Alsaad, A. K., K. J. Yousif, and M. N. AlJedaiah. 2018. Collaboration: The key to gain value from IT in supply chain. *EuroMed Journal of Business* 13, no. 2: 214–235.

Bellgran, M. 1998. Systematic design for assembly systems: Preconditions and design process planning. In *Linköping Studies in Science and Technology*, Dissertation No. 515. Linköping: Linköping University.

Bocken, N. M. P., and T. H. J. Geradts. 2020. Barriers and drivers to sustainable business model innovation: Organization design and dynamic capabilities. *Long Range Planning* 53: 1–23.

Breznik, L., M. Lahovnik, and V. Dimovski. 2019. Exploiting firm capabilities by sensing, seizing and reconfiguring capabilities: An empirical investigation. *Economic and Business Review for Central and South – Eastern Europe* 21, no. 1: 5–36.

Chanda, U., and A. Kumar. 2011. Economic order quantity model with demand influenced by dynamic innovation effect. *International Journal of Operational Research* 11, no. 2: 193–215.

Chanda, U., and A. Kumar. 2019. Optimization of EOQ model for new products under multi-stage adoption process. *International Journal of Innovation and Technology Management* 16, no. 2: 1950015.

Chong, A. Y. L., and L. Zhou. 2014. Demand chain management: Relationships between external antecedents, web-based integration and service innovation performance. *Journal of Production Economics* 154: 48–58.

Cirjevskis, A. 2019. The role of dynamic capabilities as drivers of business model innovation in mergers and acquisitions of technology-advanced firms. *Journal of Open Innovation: Technology, Market, and Complexity* 5, no. 12: 1–16.

Damanpour, F. 2010. An integration of research findings of effects of firm size and market competition on product and process innovations. *British Journal of Management* 21: 996–1010.

Ding, K., T. S. C. Felix, X. Zhang, Z. Guanghui, and Z. Fuqiang. 2019. Defining a digital twin-based cyber-physical production system for autonomous manufacturing in smart shop floors. *International Journal of Production Research* 57, no. 20: 6315–6334.

Ding, K., and P. Jiang. 2018. RFID-based production data analysis in an IoT-enabled smart job-shop. *IEEE/CAA Journal of Automatica Sinica* 5, no. 1: 128–138.

Fagerberg, J., C. D. Mowery, and R. R. Nelson, eds. 2005. *The Oxford Handbook of Innovation.* Oxford: Oxford University Press.

Ferraris, A., A. Mazzoleni, A. Devalle, and J. Couturier. 2019. Big data analytics capabilities and knowledge management: Impact on firm performance. *Management Decision* 57, no. 8: 1923–1936.

Gildea, J., and D. Foster. 2018. Agile and demand management: Giving everyone a seat at the table. *The Jabian Journal*: 51–53.

Guisado-González, M., L. T. Wright, and M. Guisado-Tato. 2017. Product—process matrix and complementarity approach. *Journal of Technology Transfer* 42, no. 3: 441–459.

Guo, Z. X., E. W. T. Ngai, C. Yang, and X. Liang. 2015. An RFID-based intelligent decision support system architecture for production monitoring and scheduling in a distributed manufacturing environment. *International Journal of Production Economics* 159: 16–28.

Haag, S., and R. Anderl. 2018. Digital twin—proof of concept. *Manufacturing Letters* 15, no. B: 64–66.

Harrison, R., J. Jaumandreu, J. Mairesse, and B. Peters. 2014. Does innovation stimulate employment? A firm-level analysis using comparable micro-data from four European countries. *International Journal of Industrial Organization* 35, no. 1: 29–43.

Helfat, E. C., and A. M. Peteraf. 2009. Understanding dynamic capabilities: Progress along a developmental path. *Strategic Organization* 7, no. 91: 91–102.

Hilletofth, P. 2012. Differentiation focused supply chain design. *Industrial Management and Data Systems* 112, no. 9: 1–18.

Hullova, D., C. D. Simms, P. Trott, and P. Laczko. 2019. Critical capabilities for effective management of complementarity between product and process innovation: Cases from the food and drink industry. *Research Policy* 48, no. 1: 339–354.

Hullova, D., P. Trott, and C. D. Simms. 2016. Uncovering the reciprocal complementarity between product and process innovation. *Research Policy* 45, no. 5: 929–940.

Jana, Paul, Stephen C. Graves, and Martin Grunow. 2018. Balancing benefits and flexibility losses in platform planning (March 4, 2018). Available at SSRN https://ssrn.com/abstract=3134037 or http://dx.doi.org/10.2139/ssrn.3134037.

Keller, T., and J. Weibler. 2015. What it takes and costs to be an ambidextrous manager: Linking leadership and cognitive strain to balancing exploration and exploitation. *Journal of Leadership and Organizational Studies* 1, no. 22: 54–71.

Kumar, A., and U. Chanda. 2016. Economic order quantity model for new product under fuzzy environment where demand follows innovation diffusion process with salvage value. *International Journal of Procurement Management* 9, no. 3: 290–309.

Kusiak, A. 2018. Smart manufacturing. *International Journal of Production Research* 56, no. 1–2: 508–517.

Lasach, O. 2019. An actor-network perspective on business models: How 'being responsible' led to incremental but pervasive change. *Long Range Plan* 52, no. 3: 406–426.

Laursen, K., and A. Salter. 2006. Open for innovation: The role of openness in explaining innovation performance among U.K. manufacturing firms. *Strategic Management Journal* 27, no. 2: 131–150.

Loch, H. C., and C. Terwiesch. 2000. Product development and concurrent engineering. *Encyclopedia of Production and Manufacturing Management* 16: 567–575.

Niewohner, N., L. Asmar, F. Wortmann, D. Roltgen, A. Kuhn, and R. Dumitrescu. 2019. Design fields of agile innovation management in small and medium sized enterprises. *Procedia 29th CIRP Design* 84: 826–831.

O´Cass, A., and L. V. Ngo. 2011. Winning through innovation and marketing: Lessons from Australia and Vietnam. *Industrial Marketing Management* 40, no. 8: 1319–1329.

Ravichandran, T. 2018. Exploring the relationships between IT competence, innovation capacity and organizational agility. *Journal of Strategic Information Systems* 27, no. 1: 22–42.

Rialti, R., L. Zollo, A. Ferraris, and I. Alon. 2019. Big data analytics capabilities and performance: Evidence from a moderated multi-mediation model. *Technological Forecasting and Social Change* 149: 119781.

Shams, R., D. Vrontis, Z. Belyaeva, A. Ferraris, and R. M. Czinkota. 2020. Strategic agility in international business: A conceptual framework for "agile" multinationals. *Journal of International Management*, doi: 10.1016/j.intman.2020.100737.

Stoyanova, V. 2018. *An analysis of David J. Teece's Dynamic Capabilites and Strategic Management: Organizing for Innovation and Growth*. (The Macat Library). Routledge, doi: 10.4324/9781912453191.

Teece, D. J. 2007. Explicating dynamic capabilities: The nature and microfoundations of (sustainable) enterprise performance. *Strategic Management Journal* 28, no. 13: 1319–1350.

Teece, D. J. 2009. *Dynamic Capabilities and Strategic Management: Organizing for Innovation and Growth*. Oxford/New York: Oxford University Press.

Teece, D. J. 2012. Dynamic capabilities: Routines versus Entrepreneurial action. *Journal of Management Studies* 49, no. 8: 1395–1401.

Tomiyama, T., E. Lutters, R. Stark, and M. Abramovici. 2019. Development capabilities for smart products. *CIRP Annals* 68, no. 2: 727–750.

Yin, D., X. Ming, and X. Zhang. 2020. Sustainable and smart product innovation ecosystem: An integrative status review and future perspectives. *Journal of Cleaner Production* 274: 1–19.

Zaman, U. K. uz, M. Rivette, A. Siadat, and S. M. Mousavi. 2018. Integrated product-process design: Material and manufacturing process selection for additive manufacturing using multi-criteria decision making. *International Journal of Robotics and Computer Integrated Manufacturing* 51: 169–180.

Zaman, U. K. uz, A. Siadat, M. Rivette, A. A. Baqai, and L. Qiao. 2017. Integrated product-process design to suggest appropriate manufacturing technology: A review. *International Journal of Advanced Manufacturing Technology* 91, no. 1–4: 1409–1430.

Zhong, Y. R., X. Xu, E. Klotz, and S. T. Newman. 2017. Intelligent manufacturing in the context of industry 4.0: A review. *Engineering* 3, no. 5: 616–630.

Zhou, G., C. Zhang, Z. Li, K. Ding, and C. Wang. 2019. Knowledge-driven twin digital manufacturing cell towards intelligent manufacturing. *International Journal of Production Research* 58, no. 4: 1034–1051.

Zhu, W., H. Tianliang, L. Weichao, Y. Yan, and Z. Chengrui. 2018. A STEP-based machining data model for autonomous process generation of intelligent CNC controller. *International Journal of Advanced Manufacturing Technology* 96, no. 1: 271–285.

Zollo, M., and S. G. Winter. 2002. Deliberate learning and the evolution of dynamic capabilities. *Organization Science* 13, no. 3: 339–351.

10 Demand Sensitivity to Time

Mohammad Kaviyani-Charati
Department of Industrial Engineering, Amirkabir
University of Technology, Tehran, Iran

Bahareh Kargar
School of Industrial Engineering, Iran University
of Science and Technology, Tehran, Iran

CONTENTS

10.1 INTRODUCTION

Today, the demand rate is more diversified over a time period due to the rapid pace of technological growth and variation in individual preferences (Kaviyani-Charati, Ghodsypour, and Hajiaghaei-Keshteli 2020; Li 2015). In various markets, such as high-tech and fashion, customer requirements and expectations change instantly, so demand calculation and estimation are challenging (Kaviyani-Charati, Ghodsypour, and Hajiaghaei-Keshteli 2020; Yang, Qi, and Li 2015). Additionally, demand for seasonal and even conventional products fluctuates because of financial incentives, shortages, and so on (Prasad and Mukherjee 2016). Thus, a key aspect of a company's success, especially in this highly competitive market, is correct demand estimation and satisfaction, which is a primary concern of supply chain managers. Inventory management is also strongly influenced by the demand that may lead to extra costs (Mishra and Singh 2011). Consequently, fit demand function development is recognized as the main challenge for most researchers and industrial practitioners (Singh, Sethy, and Nayak 2019).

Generally, demand is identified as a major component of supply chains. To be more specific, the costs of inventory control and management, procurement, transportation, customer loss, etc. have stemmed from demand (Ghosh and Chakrabarty 2009; Geetha and Uthayakumar 2009; Panda, Senapati, and Basu 2009). Furthermore, as technology and accessibility are boosted, demand is profoundly becoming uncertain and difficult to predict (Prasad and Mukherjee 2016). Thus, this intricate situation has made consumption rates variable during a sale season, and a constant function cannot be practical for real assumptions (Prasad and Mukherjee 2016; Pervin, Roy, and Weber 2018). Accordingly, better policies for procurement, fewer logistics costs, further customer satisfaction, increasing profits, and so forth could be achieved through a suitable demand function (Kumar, Singh, and Sharma 2010; Singh, Sethy, and Nayak 2019; Geetha and Uthayakumar 2009; Panda, Senapati, and Basu 2009).

The need for some products, such as crops, high-tech items, perishable products, and seasonal and fashionable goods, is deeply affected by time (Pervin, Roy, and Weber 2018; Uthayakumar and Tharani 2017; Singh et al. 2020). Besides, consumers' behaviors toward purchasing are considerably modified because of major advances in technology and Internet access that greatly affect demand (Li 2015; Hsu and Li 2006). As such, distinctive time-dependent demand functions are investigated and proposed in the literature (Li 2015). The findings have revealed that these types of functions could help decision-makers reach better results for replenishment scheduling, costs and deteriorating rates' reductions, and reducing bargains (Kumar, Singh, and Sharma 2010; Singh, Sethy, and Nayak 2019; Ghosh and Chakrabarty 2009; Geetha and Uthayakumar 2009; Panda, Senapati, and Basu 2009). Further, demand can lead not only to customer dissatisfaction but also to increasing operating costs if the production and receiving of raw materials are not well scheduled, subsequently resulting in profit and customer loss (Geetha and Uthayakumar 2009; Hsu and Li 2006; Panda, Senapati, and Basu 2009; Pervin, Roy, and Weber 2018).

In this chapter, different demand functions are surveyed for various products explored in the previous works. A research trend is also presented with promising future research directions. Besides, the applications of the functions are differently shown in the industrial examples.

10.2 CONCEPTUAL FRAMEWORK

Demand has a vital role in supply chain management that is classified into two types, namely, independent and dependent demand. Independent demand usually stems from the end-users; however, the dependent demand is derived from the independent one (Prasad and Mukherjee 2016; Pervin, Roy, and Weber 2018).

Thus, highly uncertain circumstances have made it extremely complicated for researchers and practitioners to forecast demand with unerring accuracy. More specifically, demand for food, fashion, electrical goods, and other seasonal products are significantly affected by customers' expectations and requirements and their behavior modification within a time frame (Geetha and Uthayakumar 2009; Uthayakumar and Tharani 2017). To cope, demand-related functions are well studied in the literature, especially when the products or services deteriorate over time (Uthayakumar and Tharani 2017; Singh et al. 2020; Uthayakumar and Karuppasamy 2019). Accordingly, numerous functions are developed in supply chain–related research works wherein time-dependent demand functions are grouped into four principal categories.

In this section, four time-dependent demand functions are separately discussed to specify how demand is changed over time under different situations (Panda and Saha 2010), and then a brief overview of the description is provided in Figure 10.1.

1. Linear time-dependent demand

As discussed, the constant demand function is not applicable to most inventory items. Therefore, demand is steadily either increased or decreased with time (Tripathi 2012; Pervin, Roy, and Weber 2018). Also, the function includes two main sections, constant and time-dependent parameters, with an upward or downward trend in demand per unit time (Sana and Chaudhuri 2003; Pervin, Roy, and Weber 2018). In other words, demand is uniformly modified with time, resulting from constant trade credit, cash discounts, inflation, etc. within a given period (Mishra and Sahab Singh 2011).

2. Exponential time-dependent demand

In this matter, more flexible situations are considered in the time-dependent function, which makes it fit for some inventory items. The more time that has passed, the more demand rate increases. For example, when a new product is first launched, its demand is explosively increased during a season after receiving customers' attention (Ghosh and Chakrabarty 2009; Geetha and Uthayakumar 2009). As such, its trend is exponentially boosted with time; however, the inverse effect can be seen in the same old item due to customers' changing expectations or requirements. Other situations

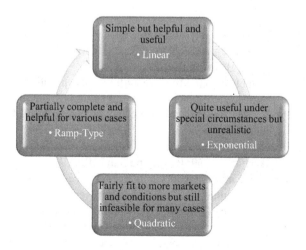

FIGURE 10.1   The brief description of the given functions.

that can lead to such an event also arise over time, for instance, providing special offers, advertising, or new fashion (Prasad and Mukherjee 2016).

3. Quadratic time-dependent demand

Sometimes the demand rate is distinctively increased or decreased; nevertheless, the rate follows a new trend. The rate is neither linear nor exponential as the accelerated and retarded increases are taken into consideration. A more realistic effect on demand rate is examined with time progress, wherein the demand has an initial value with initial change and acceleration rates. As discussed, customers can receive different offers, for example, permissible delay in payments, quantity discounts, and price discounts, leading to a higher demand rate. However, they will not be provided throughout a sale season, so the previous assumption fails to be applicable and useful for this matter. In addition, the financial incentives may be different over the period in which the demand rate will be differently affected.

4. Ramp-type time-dependent demand

In this situation, the demand rate is first increased until a certain time and then its pace slows. As time passes, the rate remains steady until it finally experiences a downward trend in general. However, this behavior of demand toward time is modified under different situations culminating with various demand functions in this area.

10.3 TIME-DEPENDENT DEMAND FUNCTIONS

As explained, numerous functions are developed in supply chain–related research works. Before progressing time-dependent demand, a noticeable amount of goods had ended up with either bargains or obsolescence (Panda and Saha 2010; Friedman 1982). In this section, different types of demand functions and their applications, related industries, and products are discussed.

Concerning time-dependent demand functions, four different types of functions are found, namely, linear, quadratic, exponential, and ramp-type (Panda and Saha 2010). They could be constructively enhancing the system's costs and productivity as the market demand has been affected with time (Li 2015; Prasad and Mukherjee 2016; Panda, Senapati, and Basu 2009).

Therefore, a well-suitable demand function will increase profits and subsequently decrease inventory obsolescence and costs when a market is highly competitive and unforeseeable (Ghosh and Chakrabarty 2009; Panda, Senapati, and Basu 2009). Hence, most researchers and practitioners are faced with the problem of choosing a well-suitable demand function to foster their responsiveness and earnings (Friedman 1982; Ghosh and Chakrabarty 2009; Panda and Saha 2010).

Simply, the demand rate in some works is linearly changed (Tripathi 2012; Pervin, Roy, and Weber 2018). By contrast, it is a highly controversial subject wherein some researchers believe demand can also be exponentially increased or decreased due to the given reasons (Friedman 1982; Ghosh and Chakrabarty 2009; Geetha and Uthayakumar 2009). In some cases, however, exponential growth is not applicable, assuming demand is steadily increased by different growth rates within a time, which is called quadratic in the literature (Khanra, Mandal, and Sarkar 2013; Singh et al. 2020). Nonetheless, the rate of demand growth or decline may be specifically accelerated, so in this situation, the ramp-type demand function is proposed (Panda, Senapati, and Basu 2009; Panda and Saha 2010).

Now, we briefly discuss the previous functions and then provide formulations in a centralized table. Considering linearity, academics have extremely debated a wide variety of models with a uniform increase or decrease in the time-dependent demand (Sana and Chaudhuri 2003; Pervin, Roy, and Weber 2018; Prasad and Mukherjee 2016). As such, variation in demand rate is assumed to be uniform with time. The demand function also includes two sections, namely, constant value and value depending on time. Figure 10.2 illustrates the trend of the linear demand rate whereby increasing and decreasing demand rate is linearly seen.

However, the main drawback of this time-dependent demand is to change uniformly per unit time. Because, the situation seldom arises in reality. In some markets, demand for novel products rises sharply while the novelty of other products wears off. So, the demand function is exponentially formulated from mathematical perspectives, and market demand for other similar products will fall drastically because of either launching a new item or changing customers' requirements (Geetha and Uthayakumar 2009). The major reason for developing such a function is that new products or crops can attract more customers; consequently, its demand will experience explosive growth and then stabilize at a saturation point (Friedman 1982). In another case, the demand for hotels and airlines are steeply boosted in holiday seasons. In this matter, demand quickly rises and falls over a time period; thus, the linear function cannot address this problem well and is formulated by the exponential demand rate (Prasad and Mukherjee 2016; Geetha and Uthayakumar 2009), which is illustrated in Figure 10.3.

Despite that, the demand rate of many items has increased during their life cycle and then fallen after introducing some new attractive products (Singh, Sethy, and Nayak 2019; Geetha and Uthayakumar 2009). Accordingly, an adverse impact on the rate owing to the negative customer impression and item deterioration has been

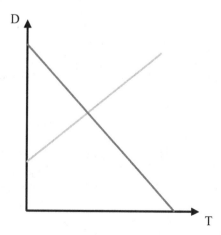

FIGURE 10.2   The behavior of demand regarding time.

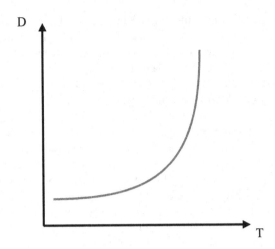

FIGURE 10.3   The behavior of demand regarding time.

found over the time frame. Item deterioration is generally defined as a value reduction or depreciation of an item during a given time period. Neither linear nor exponential functions can be appropriate since the rate is steadily increasing or accelerating within the time frame. As the previous functions are infeasible for such a situation, the quadratic demand function is proposed (Singh et al. 2020; Khanra, Mandal, and Sarkar 2013).

A quadratic function contains three different parts, namely, constant, increasing, and accelerating rates of demand, which made the formulation more practical under the situation (Khanra, Mandal, and Sarkar 2013). However, in reality, the demand rate has followed an extremely complicated pattern with time. Hence, those functions may lose their efficiency in some cases. Specifically, some items are stocked in warehouses before a sale season or at the beginning of the season. At this step, the demand rate is usually low, and its growth is not noticeable. After a certain time, the market demand is automatically increased since the need for such an item is increasingly undermined due to a decrease in the supply. Consequently, the demand rate and salvage value gradually increase with progress (Singh, Sethy, and Nayak 2019; Panda, Senapati, and Basu 2009; Sana and Chaudhuri 2003). Then, the rate has remained constant after its peak level. The ramp-type trend in quadratic time-dependent demand is depicted in Figure 10.4.

Another demand function involving shelf life and the decrease in product quality is presented. In this specific market, the product demand experiences a downward trend during its lifetime. Also in perishable items, the remaining time to the expiration date will directly affect the demand; thus, a specific function is introduced for retailing sectors to address the problem (Afshar-Nadjafi, Mashatzadeghan, and Khamseh 2016; Tripathi 2012). Perishable products should be sold to consumers before spoiling, so their features and function are different in terms of shelf life, demand volume and consumption, and perishability rate.

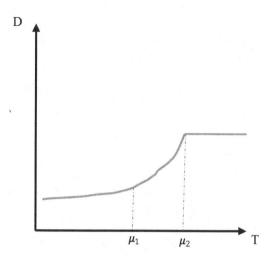

FIGURE 10.4   The behavior of demand regarding time.

Besides, Internet shopping is markedly distinct in terms of higher-order frequency and smaller-order quantity, making it intangible and costly (Hsu and Li 2006). For example, in online food shopping, peak demand may occur at lunchtime during a day. As such, serving consumers via uniform shipping cycles without any variations in cumulative quantities ordered over each shipping cycle may cause higher logistics costs and profit loss (Hsu and Li 2006). In conclusion, a quick overview of the time-dependent demand functions is provided in Table 10.1.

10.3.1 Benefits and Drawbacks

Different time-dependent demand functions are used to lower bargains, transportation costs, and inventory loss, and improve customer satisfaction and usage of natural resources. Here, the advantages and disadvantages of all given functions are briefly provided to gain a better understanding of their features.

The linear function is the simplest and most useful method to interpret demand quickly; consequently, it is an easy way to manage demand and inventory, resulting in cost reduction. However, the function cannot be applicable for various situations as the reducing or increasing rate is uniform in that sense.

In this competitive market, special offers, customer expectations and requirements, launching new products, deterioration, and so on during a sale season can cause different trends in demand rate wherein the linear function cannot be useful. Thus, exponential functions for such a situation are proposed in which special offers such as price discounts and launching novel products have heavily affected the demand rate. This function has taken exponential growth and reduction into account. Notwithstanding, the function is not only applicable to most markets but also increasing inventory costs and leading to inefficient strategies. Additionally, the rate can be neither increasing nor decreasing over the life cycle of a product. So, the

TABLE 10.1

A Short Description of the Time-Dependent Demand Functions (Authors)

Demand function	Definition	Features	References
$a \pm bt$	a and $b > 0$ are constant parameters	Simple, easy to interpret, linearly increasing or decreasing, useful for simple inventory systems	Tiwari, Wee, and Sarkar 2017
$ae^{bt} \quad 0 \leq t \leq t_1$ $a \quad t_1 \leq t \leq T$	a, $0 \leq b \leq 1$ are positive and constant, t is the length of time	Considers high fluctuation in demand and instant change in customers' expectations and is also useful for products with a high and constant demand rate	Geetha and Uthayakumar 2009
$a + bt + ct^2$	a, b, and $c > 0$ are the initial rate of demand, increasing rate of demand, and the acceleration of demand rate	Useful for a situation wherein products experience both increasing and decreasing market demand	Singh et al. 2020; Khanra, Mandal, and Sarkar 2013
$D_0 + B(t-\mu)H(T-\mu)$	D_0 is an initial demand, b is the rate of change in demand rate with time, t is time, μ is a certain time, $H(T-\mu)$ is a Heaviside function	More practical for when product demand experiences both constant conditions and slow growth considering linear trends	Sana and Chaudhuri 2003
$a_0\left[t-(t-\mu)H(t-\mu)\right]$	$a_0 \geq 0$, $\mu > 0$, $H(t-\mu)$ are the initial demand rate, a fixed point in time, and Heaviside unit step function	Useful for analyzing situations after launching newly high-tech products	Singh, Sethy, and Nayak 2019
$a + b\{t-(t-\mu)H(t-\mu)$ $+ct(t-\gamma)H(t-\gamma)\}^2$	$H(t-\mu)$ and $H(t-\gamma)$ are Heaviside functions, a is the initial rate of demand, b,c are the increasing demand rates	A more flexible and complicated function for the market wherein the new product first enters markets considering quadratic and constant changes in demand rate	Panda, Senapati, and Basu 2009

Demand function	Features	Definition	References
$A \exp\left[b\left\{t - (t-\mu)^2 H - (t-\gamma)^2 H\right\}\right]$, $0 < \mu < \gamma$	Useful for a product that has experienced comprehensive changes in demand throughout its life cycle (considering exponential increasing, constant, and decreasing trends in demand rate)	μ, γ are a fixed time, A is initial demand rate-independent time, $H(t-\mu)$ and $H(t-\gamma)$ are Heaviside functions, and b is a constant value	Panda and Saha 2010
$D_0\left[1 - \frac{t}{T}^{\omega}\right]$	Applicable for a product that has a specific expiration date	T, ω, D_0, and t are the expiration date, demand control parameter, maximum demand rate, and time	Afshar-Nadjafi, Mashatzadeghan, and Khamseh 2016
$\begin{cases}(a+bt), & 0 < t \le t_1 \\ \delta(a+bt), & t_1 < t \le T\end{cases}$	Easy to use and considering backlogging rate under linear changeable circumstances	$a, b > 0$, and δ are non-negative constants and backlogging rate	Shaikh et al. 2019

flaw in the function is addressed through the quadratic functions wherein both trends are taken into consideration.

As given, although this function is more fit for more markets, it also considers accelerated growth or fall in demand. Therefore, when the demand rate is experiencing either constant or mixed status, the mentioned functions will lose their efficiency. As such, the ramp-type time-dependent demand functions are employed to both optimize inventory and satisfy demand. In this function, all situations of a product's demand rate in its life cycle are examined and considered.

This advanced function is also comprised of linear, constant, exponential, and quadratic trends, but by contrast, it is more complicated and difficult to interpret. As mentioned, this function cannot be used for a condition whereby previous increasing or decreasing trends are just witnessed in the demand (such as linear, exponential, and quadratic). Accordingly, this function is more realistic; however, it will be time-consuming and inefficient for such situations.

10.4 INDUSTRIAL EXAMPLES

The mentioned models have been used in many different industries, such as manufacturing, food, retailing, and healthcare. In this section, two case studies are provided to support the time-dependent demand and to show how demand function changes over time.

Time-dependent demand can be seen in different inventory models. Minimizing the inventory carrying cost is the key objective of inventory management models. As such, determining the optimal time and stock of inventory replenishment to satisfy upcoming demands are of critical importance. The main goal in inventory management is to minimize the total cost associated with determining when and how many to order in the system.

Deterioration is one of the most important factors in the study of the inventory system. As previously mentioned, deterioration is referred to as harm, dissipation, waste, pilferage, outdated nature, and loss of the products (Uthayakumar and Karuppasamy 2019). Most of the items deteriorate over time. In some products, such as glassware, hardware, steel, and toys, the deterioration rate is very small. However, some products having finite shelf lives, such as medicine, blood, fish, vegetables, and food grains, deteriorate quickly over time.

Consequently, deterioration effects must be taken into consideration in inventory management while finding the optimal inventory policy for the specific product, and the storage system should be considered at the same time. In supply chain management, customer demand is crucial due to the total revenue resulting from that demand rate.

Notably, time-dependent demand can be applied in seasonal products and items, as demand typically depends on time (Sana and Chaudhuri 2003; Singh, Sethy, and Nayak 2019; Panda and Saha 2010). Notwithstanding, the stopping time of production is a key aspect in decision-making due to the perishability of seasonal products. Thus, if the production is time-dependent demand, then production stopping time and the production rate would be properly implemented. This type of demand also indicates whether the inventory is depleted at the end of the season as well as how the inventory depletion rate is affected by it. The healthcare industry is another application of a time-dependent inventory model that is of importance. The list of possible microbial ailments, chronic health diseases, and viral infections has remained remarkably consistent during the past few years. Accordingly, the demand for drugs, medicines, tablets, and other clinical products is increasing over time (Uthayakumar and Karuppasamy 2019).

Pharmaceutical inventory items are considered one of the most central resources of the healthcare industries. Pharmaceutical items are more commonly considered as medicine or drugs. As mentioned, researchers considered demand as constant in the classical inventory models. In emergencies, hospitals and other healthcare institutions need a lot of medicine, tablets, injectables, capsules, syrups, etc. Thus, the pharmaceutical demand rate cannot be constant, it varies with time (Uthayakumar and Karuppasamy 2019). The healthcare system also faces pharmaceutical shortage problems owing to defective items. In other words, the healthcare systems face a loss of goodwill or loss of profit.

The main issue comes from the fact that medicines and drugs cannot be used beyond the expiration date (Uthayakumar and Karuppasamy 2019). In pharmaceutical inventory, deterioration is a key aspect that needs to be considered. Drug expiration is one of the main concerns in public hospital/clinic pharmacies, as this occupies a huge amount of the budget to buy drugs. It should be noted that the number of medicines that expire in pharmacies can be an indication of how the medicines are used, and subsequently, it can contribute to the disease prevalence. As medicines and drugs contribute to patient care and patient life, the deterioration of pharmaceuticals should be addressed effectively.

10.5 MATHEMATICAL MODELS

Two instances of mathematical models used to formulate related practical studies in the literature are presented as follows.

10.5.1 First Model

A mathematical modeling for a distributor's delivery strategy problem is proposed, considering carbon emissions, demand–supply interactions, and time-dependent demand of retailers (Li 2015). The proposed mathematical model aims to obtain the optimal number and time window of service cycles. The detailed description is shown as follows:

Terminology and Notations

i	The number of delivery service cycles	$\tilde{W}_k$	The weight capacity of vehicle type k
s	The specific delivery service cycle	t_i^0	The start time of the service cycle i
k	The specific vehicle type	t_i^m	The end time of the service cycle i

p	Product	π	The total profit of the distributor
T_i	The time window of the service cycle i, $T_i = \left(t_i^0, t_i^m\right)$	$L_{i,k}$	The delivery cost of vehicle type k of service cycle i
b_p	The distributor's acquisition cost of product p	$c(T, T_1)$	The total cost per year

Decision Variable

$\delta_{i,k}$ If vehicle type k is assigned to service cycle i, then $\delta_{i,k}=1$; otherwise, $\delta_{i,k}=0$

Model Development
In this study, Li (2015) assumed that the retailer could get price discounts from the distributor through the delays in receiving the products that were ordered. Therefore, two scenarios including with and without a price discount are investigated. The distributor's total profit without a price discount can be formulated as follows:

$$\max \pi = \sum_{i=1}^{S}(\sum_{\forall p}(\widetilde{b_p}-b_p)Q_i^p)-\sum_{i=1}^{S}L_{i,k} \tag{10.1}$$

$$(\delta_{i,k}\tilde{W}_k - \sum_{t=t_i^0}^{t_i^m}\sum_{\forall p}Q_t^pW_p)\geq 0, \quad \forall i \tag{10.2}$$

$$\sum_{i=1}^{S}(t_i^m - t_i^0) = T \tag{10.3}$$

$$t_{i+1}^m = t_i^0, \quad i=1,2,3,....(S-1) \tag{10.4}$$

$$t_s^m = t_1^0 \tag{10.5}$$

$$t_i^m > 0, \quad \forall i \tag{10.6}$$

$$t_i^0 > 0, \quad \forall i \tag{10.7}$$

The distributor's total profit with a price discount is as follows:

$$\max \pi = \sum_{i=1}^{S}(\sum_{\forall p}\widetilde{b_p}(1-\theta)-b_p)Q_i^p)-\sum_{i=1}^{S}L_{i,k} \tag{10.8}$$

Solution Approach
An exact method is applied to find optimal solutions. To do so, the proposed model is applied to a logistics case study in Taiwan to find optimal delivery strategies for the distributor, considering the carbon-pricing mechanism.

10.5.2 SECOND MODEL

Shukla, Shukla, and Yadava (2013) proposed an inventory model for deteriorating items considering shortages, which is partially backlogged with the aim of total cost minimization. In other words, the shortages are allowed wherein a part of the unsatisfied demand will be postponed being met later. In their study, the demand is considered an exponential time-dependent demand rate.

Terminology and Notations

T	Cycle time	T_1	The time at which $I(t)=0$
$I(t)$	The inventory level at the time t	A	Set of cost
$\lambda_0 e^{\alpha t}$	Demand rate of items	$C(T, T_1)$	The total cost per year
θ	The constant deterioration rate	c_s	The shortage cost per unit of time
T_i	The time window of the service cycle i, $T_i = \left(t_i^0, t_i^m\right)$	$L_{i,k}$	The delivery cost of vehicle type k of service cycle i
c_d	The cost of each deteriorated item	c_h	The inventory holding cost per unit of time

Model Development

The inventory level varies over time, owing to both demand and deterioration of the material. The total cost including deterioration, shortage, and holding cost is as follows:

$$\min C(T_1,T) = \frac{1}{T}\left[\frac{c_h\lambda_0}{(\alpha+\theta)}\left\{\frac{e^{\alpha T_1}}{\theta}\left(e^{\theta T_1}-1\right)-\frac{1}{\alpha}\left(e^{\alpha T_1}-1\right)\right\}\right.$$
$$\left.+c_d\lambda_0\left\{\frac{1}{(\alpha+\theta)}\left(e^{(\alpha+\theta)T_1}-1\right)-\frac{1}{\alpha}\left(e^{\alpha T_1}-1\right)\right\}+\frac{c_s\lambda_0}{2}(T-T_1)^2+2A\right] \quad (10.9)$$

Solution Approach

To find an optimal solution in exponential terms, the truncated Taylor's series is applied. We get Eq. (10.10) by using a truncated Taylor's series expansion in Eq. (10.9).

$$C(T_1,T) = \left[\frac{\lambda_0(c_h+\theta c_d+c_s)\,T_1^2}{2T}+\frac{c_s\lambda_0 T}{2}-c_s\lambda_0 T_1+\frac{2A}{T}\right] \quad (10.10)$$

By taking the first and second partial derivatives of Eq. (10.10) concerning T and T_1, the optimal solution is achieved. Finally, by solving simultaneously

$\dfrac{\delta C\left(T_1, T\right)}{\delta T_1} = 0$ and $\dfrac{\delta C\left(T_1, T\right)}{\delta T} = 0$, the minimum optimal solution of $T = T^*$ and $T_1 = T_1^*$ is achieved as follows:

$$\lambda_0\left(c_h + \theta c_d + c_s\right)T_1 - c_s T = 0 \tag{10.11}$$

$$\lambda_0\left(c_h + \theta c_d + c_s\right)T_1^2 - \lambda_0 c_s T^2 + 4A = 0 \tag{10.12}$$

10.6 RESEARCH TRENDS

In this section, previous studies related to time-dependent demand are classified and surveyed. Investigation of recent trends in the work will open the door for further research on this area, leading to novelties and significant contributions to the subject area.

The constant demand is initially considered in the supply chain; however, the assumption fails to be fit for this competitive market. In reality, demand can be changed with space, inflation, income, customer expectation and requirements, *etc.* For example, sales of seasonal items like fashionable clothes, domestic goods, electronic products, and food substantially grow when attracting more consumers. Thus, sales for other products can considerably decline due to a marked shift in consumer preferences and also the launching of cutting-edge items (Geetha and Uthayakumar 2009; Uthayakumar and Tharani 2017; Singh et al. 2020; Uthayakumar and Karuppasamy 2019). In this situation, the demand rate will be modified and reformulated through time-varying functions (Panda and Saha 2010).

Therefore, different types of inventory models considering the time-varying demand have been given attention. Regarding that, a time-dependent demand was initially proposed by Silver and Meal (1969). First, the time-dependent demand is developed using a linear equation. Then, Bose, Goswami, and Chaudhuri (1995) presented a deterministic inventory model for an item with a linear pattern in demand. In another paper, Donaldson (1977) analyzed an analytical solution for the inventory replenishment problem considering linear time-dependent demand over the finite-time horizon. This kind of continuous linear function of demand was applied in other papers (Sana, Chaudhuri, and Mahavidyalaya 2004; Kumar, Singh, and Sharma 2010; Malik, Singh, and Gupta 2008; Shaikh et al. 2019).

Nevertheless, the formulation has been gradually evolved into a more fit function, namely, an exponential one. Thus, several models have been optimized using the exponential time-dependent demand function. As stated, demand for newly launched products may increase quickly as they come to markets (Friedman 1982; Ghosh and Chakrabarty 2009; Geetha and Uthayakumar 2009; Hariga and Benkherouf 1994).

Inversely, demand fluctuation in the real situation cannot be exponential, as its change is extraordinarily high and exceptional (Khanra, Mandal, and Sarkar 2013; Singh et al. 2020). To overcome that, Khanra and Chaudhuri (2003) examined its weaknesses wherein they proved that exponential demand function cannot be practical under different circumstances. As such, a quadratic function considering accelerated growth in the demand rate is presented, so the demand rate is gradually changed compared to the exponential demand function (Khanra, Mandal, and Sarkar 2013; Singh et al. 2020). Notwithstanding, the accelerated decline or growth in demand cannot be continued forever (Panda, Senapati, and Basu 2009; Sana and Chaudhuri 2003; Panda and Saha 2010). Thereafter, a realistic approach is to formulate it as three components of ramp-type time-dependent function.

The ramp-type demand function is more practical in analyzing situations for newly launched high-tech products (Singh, Sethy, and Nayak 2019; Sana and Chaudhuri 2003). This model can be found in Mandal and Pal (1998), Wu and Ouyang (2000), Manna and Chaudhuri (2006), and Singh, Sethy, and Nayak (2019). Hill (1995) firstly proposed a ramp-type time-dependent function where the demand has two different types of demand patterns in two successive periods during the period. For outlining the promising research directions, a brief overview of the research trend in time-dependent demand functions is given in Table 10.2.

10.6.1 FUTURE RESEARCH DIRECTIONS

There is a growing body of literature that recognizes the importance of time-dependent demand and its applications. In the literature, the relative significance of deterioration rate, shortages, and backlogging have been subject to considerable discussion. However, the research on the subject has been mostly restricted to deterministic circumstances, so most effective rates are considered a constant value. Accordingly, further research on this area could be devoted to uncertain deterioration rates and demand, as this feature will provide more reliable and practical implications for policymakers to enhance their business. Also, recent research has highlighted the need for a time-dependent deterioration rate, while a limited number of research has been conducted in this subject area. So, this matter would be another promising future research direction. Last, various trade credits, for instance, partial and two-level trade credit, have not been fully understood and investigated in the literature, which can make them an interesting research area to illuminate useful knowledge and managerial implications for decision-makers.

10.7 CONCLUSION

Nowadays, as the market is highly competitive and unforeseeable, a well-suitable demand function will increase profits and subsequently decrease inventory obsolescence and costs. Therefore, most researchers and practitioners are faced with problems of choosing a fit demand function to foster their responsiveness and earnings. Demand rates change with time when it comes to the market. Besides the constant demand, some other demand functions that can formulate the actual behavior of demand more precisely are introduced. This chapter focuses on time-dependent demand and various proposed demand functions. Concerning time-dependent functions, four different types of demand are found based on the literature, namely, linear, quadratic, exponential, and ramp-type. Moreover, the research trend is provided to get a brief overview of time-dependent demand, and future research directions are outlined.

TABLE 10.2
The Quick Overview of the Literature Review in Time-Dependent Demand Functions

References	SC level	Product features	Objective function	Demand function	Solution approach
Afshar-Nadjafi, Mashatzadeghan, and Khamseh 2016	Retailing	Perishable items with a rigid expiration date	Minimizing costs	Descending demand rate during the products' lifetime	Optimization solver in the Matlab
Ghosh and Chakrabarty 2009	Storage and retailing	Deteriorating items	Minimizing inventory and storage costs	Exponential demand rate	Newton–Raphson approach
Hariga and Benkherouf 1994	Inventory replenishment management	Deteriorating products	Optimizing replenishment schedule	Exponential growth	Heuristic approaches
Sudhansu and Chaudhuri 2003	Retailing or manufacturing	Computer and computer-based	Minimizing inventory costs	Quadratic demand rate	Heuristic approach
Li 2015	Distribution management	Various products	Minimizing green and logistics costs	Discrete time-dependent demand	Visual C++
Panda and Saha 2010	Manufacturing	Perishable items	Maximizing profit	Exponential ramp-type	Differentiating procedure
Pervin, Roy, and Weber 2018	Retailing industries	Deteriorating products	Minimizing costs	Linear demand	
Geetha and Uthayakumar 2009	Retailing	Non-instantaneous deteriorating items	Minimizing costs	Exponential and constant demand rate	
Khanra, Mandal, and Sarkar 2013	Retailing industries	Seasonal and technological products	Minimizing costs	Quadratic demand rate	
Kumar, Singh, and Sharma 2010	Manufacturing	Deteriorating items	Maximizing profit	Linear demand rate	
Malik, Singh, and Gupta 2008	Manufacturing	Deterioration items	Minimizing costs	Linear demand	

References	SC level	Product features	Objective function	Demand function	Solution approach
Manna and Chaudhuri 2006	Manufacturing systems	Deteriorating items	Minimizing inventory costs	Ramp-type demand	
Mishra and Singh 2011	Retailing and manufacturing	Deteriorating items	Minimizing total inventory costs	Linear demand	
Singh et al. 2020	Retailing and manufacturing	Deteriorating items	Minimizing logistics costs	Quadratic demand	
Singh, Sethy, and Nayak 2019	Retailing industries	High-tech products	Boosting profits	Ramp-type demand	
Uthayakumar and Karuppasamy 2019	Retailing management	Pharmaceutical items	Minimizing costs	Exponential demand	
Wu and Ouyang 2000	Manufacturing or retailing systems	Deteriorating items	Minimizing costs	Ramp-type demand	
Sana and Chaudhuri 2003	Production systems	Deteriorating items	Maximizing profit	Ramp-type demand	Newton–Raphson method
Shaikh et al. 2019	Retailing systems	Deteriorating items	Boosting profits	Linear demand	Matlab software
Panda, Senapati, and Basu 2009	Manufacturing and retailing industries	Deterioration items	Minimizing costs and maximizing profit	Quadratic ramp-type demand	Newton–Raphson method
Tiwari, Wee, and Sarkar 2017	Retailing	Deteriorating items	Maximizing profit	Ramp-type demand	Heuristic approach
Tripathi 2012	Retailing industries	Deteriorating items	Minimizing costs	Linear demand	Taylor's series
Uthayakumar and Tharani 2017	Production management	Deteriorating items	Minimizing costs	Exponential demand	Iterative solution method

REFERENCES

Afshar-Nadjafi, B., H. Mashatzadeghan, and A. Khamseh. 2016. Time-dependent demand and utility-sensitive sale price in a retailing system. *Journal of Retailing and Consumer Services* 32: 171–174.

Bose, S., A. Goswami, and K. S. Chaudhuri. 1995. An EOQ model for deteriorating items with linear time-dependent demand rate and shortages under inflation and time discounting. *Journal of the Operational Research Society* 46, no. 6: 771–782.

Donaldson, W. A. 1977. Inventory replenishment policy for a linear trend in demand—an analytical solution. *Journal of the Operational Research Society* 28, no. 3: 663–670.

Friedman, M. 1982. Inventory lot-size models with general time-dependent demand and carrying cost functions. *INFOR: Information Systems and Operational Research* 20, no. 2: 157–167.

Geetha, K. V., and R. Uthayakumar. 2009. Optimal inventory control policy for items with time-dependent demand. *American Journal of Mathematical and Management Sciences* 29, no. 3–4: 457–476.

Ghosh, S., and T. Chakrabarty. 2009. An order-level inventory model under two level storage system with time-dependent demand. *Opsearch* 46, no. 3: 335.

Hariga, M. A., and L. Benkherouf. 1994. Optimal and heuristic inventory replenishment models for deteriorating items with exponential time-varying demand. *European Journal of Operational Research* 79, no. 1: 123–137.

Hill, R. M. 1995. Inventory models for increasing demand followed by level demand. *Journal of the Operational Research Society* 46, no. 10: 1250–1259.

Hsu, C.-I., and H.-C. Li. 2006. Optimal delivery service strategy for Internet shopping with time-dependent consumer demand. *Transportation Research Part E: Logistics and Transportation Review* 42, no. 6: 473–497.

Kaviyani-Charati, M., S. H. Ghodsypour, and M. Hajiaghaei-Keshteli. 2020. Impact of adopting quick response and agility on supply chain competition with strategic customer behavior. *Scientia Iranica.*

Khanra, S., and K. S. Chaudhuri. 2003. A note on an order-level inventory model for a deteriorating item with time-dependent quadratic demand. *Computers and Operations Research* 30, no. 12: 1901–1916.

Khanra, S., B. Mandal, and B. Sarkar. 2013. An inventory model with time dependent demand and shortages under trade credit policy. *Economic Modelling* 35: 349–355.

Kumar, V., S. R. Singh, and S. Sharma. 2010. Profit maximization production inventory models with time dependent demand and partial backlogging. *International Journal of Operational Research Optimization* 1, no. 2: 367–375.

Li, H.-C. 2015. Optimal delivery strategies considering carbon emissions, time-dependent demands and demand—supply interactions. *European Journal of Operational Research* 241, no. 3: 739–748.

Malik, A. K., S. R. Singh, and C. B. Gupta. 2008. An inventory model for deteriorating items under FIFO dispatching policy with two warehouse and time dependent demand. *Ganita Sandesh* 22, no. 1: 47–62.

Mandal, B., and A. K. Pal. 1998. Order level inventory system with ramp type demand rate for deteriorating items. *Journal of Interdisciplinary Mathematics* 1, no. 1: 49–66.

Manna, S. K., and K. S. Chaudhuri. 2006. An EOQ model with ramp type demand rate, time dependent deterioration rate, unit production cost and shortages. *European Journal of Operational Research* 171, no. 2: 557–566.

Mishra, V. K., and L. Sahab Singh. 2011. Deteriorating inventory model for time dependent demand and holding cost with partial backlogging. *International Journal of Management Science and Engineering Management* 6, no. 4: 267–271.

Panda, S., and S. Saha. 2010. Optimal production rate and production stopping time for perishable seasonal products with ramp-type time-dependent demand. *International Journal of Mathematics in Operational Research* 2, no. 6: 657–673.

Panda, S., S. Senapati, and M. Basu. 2009. A single cycle perishable inventory model with time dependent quadratic ramp-type demand and partial backlogging. *International Journal of Operational Research* 5, no. 1: 110–129.

Pervin, M., S. K. Roy, and G.-W. Weber. 2018. Analysis of inventory control model with shortage under time-dependent demand and time-varying holding cost including stochastic deterioration. *Annals of Operations Research* 260, no. 1–2: 437–460.

Prasad, K., and B. Mukherjee. 2016. Optimal inventory model under stock and time dependent demand for time varying deterioration rate with shortages. *Annals of Operations Research* 243, no. 1–2: 323–334.

Sana, S., and K. S. Chaudhuri. 2003. An EOQ model with time-dependent demand, inflation and money value for a ware-house enterpriser. *Advanced Modeling and Optimization* 5, no. 2: 135–146.

Sana, S., K. S. Chaudhuri, and B. Mahavidyalaya. 2004. On a volume flexible production policy for a deteriorating item with time-dependent demand and shortages. *Advanced Modeling and Optimization* 6, no. 1: 57–74.

Shaikh, A. A., G. C. Panda, S. Sahu, and A. K. Das. 2019. Economic order quantity model for deteriorating item with preservation technology in time dependent demand with partial backlogging and trade credit. *International Journal of Logistics Systems and Management* 32, no. 1: 1–24.

Shukla, H., V. Shukla, and S. Yadava. 2013. EOQ model for deteriorating items with exponential demand rate and shortages. *Uncertain Supply Chain Management* 1, no. 2: 67–76.

Silver, E. A., and H. C. Meal. 1969. A simple modification of the EOQ for the case of a varying demand rate. *Production and Inventory Management* 10, no. 4: 52–65.

Singh, T., M. M. Muduly, N. Asmita, C. Mallick, and H. Pattanayak. 2020. A note on an economic order quantity model with time-dependent demand, three-parameter Weibull distribution deterioration and permissible delay in payment. *Journal of Statistics and Management Systems*: 1–20.

Singh, T., N. N. Sethy, and A. K. Nayak. 2019. An ordering policy with generalized deterioration, ramp-type demand under complete backlogging. In *Logistics, Supply Chain and Financial Predictive Analytics*, 43–55. Singapore: Springer.

Tiwari, S., H.-M. Wee, and S. Sarkar. 2017. Lot-sizing policies for defective and deteriorating items with time-dependent demand and trade credit. *European Journal of Industrial Engineering* 11, no. 5: 683–703.

Tripathi, R. P. 2012. EOQ model for deteriorating items with linear time dependent demand rate under permissible delay in payments. *International Journal of Operations Research* 9, no. 1: 1–11.

Uthayakumar, R., and S. K. Karuppasamy. 2019. An EOQ model for deteriorating items with different types of time-varying demand in healthcare industries. *Journal of Analysis* 27, no. 1: 3–18.

Uthayakumar, R., and S. Tharani. 2017. An economic production model for deteriorating items and time dependent demand with rework and multiple production setups. *Journal of Industrial Engineering International* 13, no. 4: 499–512.

Wu, K.-S., and L. Y. Ouyang. 2000. A replenishment policy for deteriorating items with ramp type demand rate. *Proceedings of the National Science Council Republic of China Part A: Physical Science and Engineering* 24, no. 4: 279–286.

Yang, D., Qi, E., and Y. Li. 2015. Quick response and supply chain structure with strategic consumers. *Omega* 52: 1–14.

11 Inflation-Dependent Demand

Abdollah Arasteh
Babol Noshirvani University of Technology

CONTENTS

11.1 INTRODUCTION

Inventory is a commodity that is stored for some time. In short, it includes goods, materials, and parts that are used in the production, sale, and management of an industry. The most complete and comprehensive definition that can be provided in terms of inventory includes items of tangible assets belonging to an enterprise, which is maintained or produced for sale in the ordinary course of business.

Inflation is seen as increasing the price of a unit of produced goods in the lack of effective control. Inflation is usually associated with a real or potential increase in the general level of price. Until the 1970s, models of inventory control assumed that the cost was fixed during the planning period. Since the early 1990s, inflationary inventory models have been developed under more dynamic and complex conditions. Among the items considered so far in inflation inventory models are: (1) demand rate is fixed and well known, (2) demand is the linear function of time, (3) demand is the function of initial inventory, and (4) demand is price-dependent.

The purpose of this chapter is to study the effect of inflation on demand from an operations management perspective. Therefore, in this section, we first refer to different strategies for responding to demand in supply chains in general. In manufacturing companies, goods are stored in warehouses and other places after production, which complicates the supply chain. If companies use a make-to-order manufacturing strategy, there will be no need to stock manufactured products, but at the same time, there will still be a need to stock raw materials and components. Therefore, supply chains are definitely dependent on the nature and status of companies. The following strategies are commonly used in supply chain demand management:

1. Integrated make to stock: Focuses on real-time customer demand to effectively store inventory.
2. Continuous storage: Based on refilling depleted inventory by working closely with suppliers or intermediaries.
3. Make to order (MTO): Based on the order to assemble the product immediately after receiving the order.
4. Engineering to order (ETO): Emphasizes product design after the customer declares the need.
5. Assembly to order: The core of the assembly is the same for most products and only changes to other final assembly components are possible.
6. Channel assembly: With a slight change in the MTO model, the parts of each product are collected and assembled as they move in the distribution channel.

Regardless of the different strategies available to meet demand in supply chains, the impact of inflation on demand in each case is significant. The relationship between inflation and demand is complex and, in some cases, unpredictable. In general, we have two types of demand for types of goods (and services): consumer demand and non-consumer (investment) demand. Consumer demand involves the demand of individuals or organizations that need to buy goods for their own consumption. Non-consumer demand includes individuals or organizations that are either looking for investment or business. Investors buy goods for the purpose of long-term investment and generally stockpile them after purchase. The goal of this group is to gain returns and maintain purchasing power against inflation in the long run. Rising inflation in the community or significant and unusually high inflation expectations in the near future will encourage the purchase of goods by this group. Another group that buys goods in a non-consumer group is traders. The trader buys a variety of assets with the aim of making a profit from price fluctuations and with a short-term goal and does not seek ancillary returns from the asset, such as renting property or dividends (unlike the investor).

11.2 CONCEPTUAL FRAMEWORK

Inventory can be characterized as the storage of a product that will be used to fulfill any future need. The product may be raw material, the pieces imported, etc. In this section, a schematic overview of the inventory structure is considered. Classic models of economic order quantity (EOQ) and economic production quantity (EPQ) are widely used in inventory control. Demand plays the role of driving force in the whole inventory system. In studying inventory, demand is a key factor that should be considered.

Wee and Law (2001) proposed an inventory model for perishable items in which time value for money is present and demand is price dependent. Papachristos and Skouri (2003) studied the model in which demand is a function of the selling price, and the back-order rate is a function of time. Both considered the corruption rate of Weibull distribution with two parameters. Balkhi and Benkherouf (2004), Pal, Bhunia, and Mukherjee (2006), and Hsu, Wee, and Teng (2007) combined various factors and examined the effect of combining these factors on demand. In their models, the optimal inventory level theorem has been considered. Saxena et al. (2020) developed an inventory model in which demand was influenced by time and inflation.

11.2.1 REVIEW OF MODELS OF INFLATION OF ORDINARY ITEMS

Inflation can be divided into two categories based on its origin.

Demand pull inflation occurs when the increase in the level of demand relative to the supply possibilities is unbalanced. Increased demand can come from the real sector of the economy, its monetary sector, or a combination of the two. Although increasing private-sector demand can also be effective, because households and businesses can only increase aggregate demand by withdrawing savings or borrowing, this increase in demand is unlikely to be significant. But governments can have important inflationary pressures to cover their expenses, such as the power to collect taxes or issue banknotes, and to obtain loans and credits from the domestic private sector, banks, or abroad. Among these methods, tax collection has the least impact and banknote issuance has the most impact. But explaining the mechanism of these effects is beyond the scope of this article.

Cost push inflation occurs when the holders of the factors of production demand a larger share of total output, which is either due to the increase in wage-cost pressure or to the profit pressure. Some anti-capitalist economists see profit pressure as a factor in rising prices and, in fact, see wage increases as a defensive move by unions. On the other hand, the proponents of capitalism say that profit is a small part of the price, and therefore the pressure to increase profits is not an important factor in increasing prices, but the rapid increase in wages versus the growth of labor productivity is the most important factor in raising prices.

In this section, we will look at the models proposed in the research literature in the field of combining traditional inventory models, such as EOQ and EPQ with inflation. Sarker and Pan (1994) studied the incomplete production process with variable demand compared to the time and considering inflation for the economic production model. Ben-Daya and Zamin (2002) presented an integrated inventory model considering the inflation rate while demand is probabilistic and delivery time is controllable. Hou and Lin (2006) examined the EOQ model for perishable items at price-dependent selling rates and inventories, taking into account the time value of money. Thangam and Uthayakumar (2010) introduced the inventory model of perishable items with inflation-related demand and overdue orders. Valliathal and Uthayakumar (2010) investigated the issue of production for perishable and improving items at the expense of nonlinear shortages under inflation and time discounts. Maity and Maiti (2008) proposed an inflation-related demand rate and storage under inflation and the time value of money. Taheri Tolgari, Mirzazadeh, and Jolai (2012) examined the inventory model for defective items in inflation conditions and assumption of review errors. Gilding (2014) presented inflation in the planning of the optimum inventory for a limited horizon.

Muniappan, Uthayakumar, and Ganesh (2015) developed an economically optimal ordering model for perishable items, taking into account inflation and the time value of money and time-dependent volatility rates and time-delayed payment delays. In their research, the shortage is allowed and is considered as a decline, and a fixed credit period is provided by the supplier to the retailer. Pareek and Dhaka (2015) presented a fuzzy inventory control model by determining the optimal ordering amount for perishable items using the cash flow discount method when the supply credits are related to the order quantity. Yang, Lee, and Zhang (2013) presented a model inventory with perishable products with demand dependent on inventory level and inflation and credit payment policy. The length of time allowed for the delay in payment by the supplier to the retailer depends on the amount of the retailer's order. Motallebi and Zandieh (2017) presented the order to locate the customer's point of influence and determine the optimal point of the order.

11.3 LIST OF DEMAND FUNCTIONS

Regarding inflation-related demand, it should be noted that the existing models in this area can be divided into four general categories.

1. Inventory models that consider inflation and time value of money with a steady demand rate (static models)
2. Inventory models of different time criteria (dynamic models)
3. Inventory models that consider the impact of inflation and time value of money with differing demand rates for time
4. Inventory models with allowable delays in payment

Table 11.1 summarizes examples of mathematical models in each category and the characteristics of each.

11.4 SUPPLY CHAIN ECHELONS WHERE THE FUNCTIONS ARE MORE APPLICABLE

Today, with the growing sophistication of production methods and the need to create a range of products and services, one company alone is no longer be able to manufacture goods or offer adequate services without the help and collaboration of other organizations.

Therefore, it is necessary to pay attention to the supply chain as the main structure of the principle of competitiveness in today's world. A supply chain is a network of materials, information, and services that are processed in relation to supply, transmission, and demand characteristics. The concept of supply chain, from one perspective, derives from how organizations participate in a particular chain that are interconnected, and its components are shown in Figure 11.1 as the SCM house model.

One of the most important components and elements of supply chain management is coordination. Coordination of information, materials, and financial flows as the

TABLE 11.1

Classification of Inflation-Dependent Demand Functions in Inventory Literature

Demand function	Parameters, variables, and conditions	Reference
$D(i_1, i_2) = a + bi_1 + ci_2$	i_m: The internal (for $m=1$) and external (for $m=2$) inflation rates; $a > 0$, b, $c < 0$	(Mirzazadeh, Seyyed Esfahani, and Fatemi Ghomi 2009)
$\tilde{D}_i(t) = a_i + b_i(\tilde{a} + bt) = \tilde{A}_i + B_i(t)$	Imprecise demand rate per unit time, which is a linear function of the inflation rate $(= a_i + b_i\tilde{\rho}, a_i > 0, b_i < 0)$ $\tilde{A}_i = a_i + b_i\tilde{a} = a_i + b_i(a_{i1}, a_{i2}, a_{i3})$ $= (A_{i1}, A_{i2}, A_{i3})$; $B_i = b_i b$	(Pathak and Sarkar 2010)
$D(p,t) = (a - bp)e^{\lambda t}$ $D(p,t) = ap^{-b}e^{\alpha t}$	$a, b > 0$; D is a linear function of the price and decreases (increases) exponentially with time if $\lambda < 0$ ($\lambda > 0$), respectively. p: price t: time α: constant inflation rate	(Ghoreishi, Mirzazadeh, and Weber 2014) (Tripathi 2015)
$f(t) = ae^{\lambda t}$	a, λ are constants such that $a > 0$, $\lambda > 0$. at $t=0, f(t)=a$	(Yadav et al. 2017)
$D(t) = (a - bp^t)$	$t \geq 0$, $a > b > 0$ and $0 < \rho < 1$	(Manna 2019)
$D_t(t) = \dfrac{\rho}{(mp_0 e^{-(i-1)RT})\gamma}(a - bt)$	The base demand of the item depends on time, promotional effort, and inflation-based selling price. ρ is the retailer's promotional effort, m is a decision variable, p_0 is the selling price at time zero, $R=d-I$, where d is the discount rate and I is the inflation rate, a, b, (≥ 1) are the parameters chosen to best fit the demand function.	(Pramanik and Maiti 2019)
$D(t) = ae^{bt}$	Demand rate is exponentially increasing, where $0 \leq b \leq 1$ is a constant inflation rate, $a > 0$.	(Yadav and Swami 2019)
$D(A_C, p) = A_C^\eta a p^{-b}$	The demand rate D is a deterministic function of selling price (p) and advertisement cost (A_C) per unit item, where $a > 0$, $b > 1$, $0 \leq \eta < 1$, a is the scaling factor, b is the index of price elasticity, and η is the shape parameter.	(Udayakumar, Geetha, and Sana 2020)

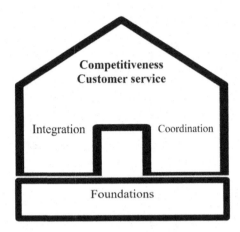

FIGURE 11.1 House of SCM (adapted from Stadtler and Kilger 2002).

second main component of supply chain management (after integration) has three parts as follows:

1. Use of information and communication technology
2. Process orientation
3. Advanced planning

Regarding the effect of inflation on different types of communications in the supply chain, it should be said that inflation affects both B2B and B2C communications. This is because, in inflationary conditions, the company has to buy its goods and raw materials at a higher price than the supply companies and also eventually sell the goods to the consumer at a higher price. Therefore, coordination and how to create it in supply chains in the face of inflation is an important issue that must be considered. Finally, the cost price of the product should not be too affected by inflation, and the manufacturer should not manage cost price in a way that exceeds the customer's capacity because it will directly affect sales and sometimes even the durability and survival of the company.

Inflation induces a rise in selling price and a reduction in consumer demand, which result from seasonal product management and occur over the long production lead period. As an important method for hedging against uncertainties, option contracts, including call, put, and bidirectional option contracts, have been proven to favor two participants of a one-supplier and one-retailer supply chain under inflation.

Several factors impact the profitability of a supply chain. The advantages of the hedge network are influenced by the effect of these factors on the volume of revenue. Inflation rate is known to be a factor that affects the earnings of producers. Raw materials, labor rates, shipping, servicing, and many other products are subject to adjustments. Multi-level inventory structures can be divided into serial, assembly, delivery, and general networks. The approach to inventory control has a significant effect on the average amount of inventories in the supply chain. In Table 11.2, we classified several models that were formulated by a significant number of researchers.

TABLE 11.2

Classification of Multi-Echelon Inventory Models

Criteria	Classification
Demand	Stochastic, deterministic, static, emergency deliveries
Inventory	Continuous review policy, periodic review policy, centralized replenishment decisions, decentralized replenishment decisions
Product	Single, multiple
# of echelons	2–3, >3

11.5 INDUSTRIES AND EXAMPLES

In today's global economy, fierce competition between companies has led them to operate in conditions of uncertainty, resulting in high risks. Risks have negative effects on the supply chain of companies and can lead to reduced profitability and competitive advantage.

Industries that have been in the realm of inflation-dependent demand, that is, some form of supply side demand in their supply chain, have been affected by existing inflation, starting with oil and gas and mining companies, then water and electricity, biotechnology, and pharmacy, and now has spread to nearly all industries, particularly communications and high-tech industries.

Few industries and supply chains can be found that are not affected by inflation-dependent demand. In the past, pharmaceutical supply chains were used as a mechanism for supplying drugs to consumers. Recently, pharmaceutical companies are searching for new ways to generate extra value. The key challenges the pharmaceutical supply chain industry face are the need to align potential ability with projected demand at the planned level and maintaining responsiveness at the operating stage. The interdependencies between these components make it necessary to implement sophisticated supply chain optimization techniques. Next are some examples of how inflation has affected different industries.

Automotive industry: As an example, General Motors (GM) adjusted the production run of its new series of cars to the inflation of the automotive industry. In this way, GM retained more raw materials, and multiple global retailers encountered surpluses for similar materials with contractual obligations that went beyond what is necessary.

Computer industry: In the computer industry, HP-Compaq used inflation and its impact on demand to predict printer sales in foreign countries. Extraordinarily special printers were then configured, assembled, and shipped to these countries. However, knowing that demand was changing rapidly and forecast figures are seldom correct, pre-configured printers often suffer from the cost of maintaining more material inventory or the cost of technology obsolescence.

Oil and gas industry: In the oil and gas industry, companies spend millions of dollars on modernizing their refineries and installing new technology to counteract the effects of inflation on demand to change the composition of their emissions from heating oil, diesel, and other petrochemicals into products.

11.6 EXAMPLE OF OPTIMIZATION MODELS

Integrated inventory model considering the time value of money and inflation

In this chapter, we develop an integrated inventory model for durable goods in a two-level supply chain. The symbols used are:

n Number of shipments

m Raw material consumption coefficient

z Difference between time value of money and inflation

D Annual demand rate

R Seller's annual production rate ($R \geq D$)

S Cost of setting up a vendor

A Buyer ordering fee

O The cost of ordering raw materials from the seller

r Annual holding rate

C_R The cost of purchasing each unit of raw material

C_v Cost per unit of the seller's final product

C_b Cost of purchasing each unit of the final product that the buyer pays to the seller ($C_b > C_v > C_R$)

π Annual cost per buyer shortage unit

F Fixed shipping cost

v Variable shipping costs

b The amount of buyer shortage per shipment

B The amount of buyer shortage per cycle ($B = nb$)

Development of mathematical models

The raw material cost function for the seller, whose inventory system behavior is shown in Figure 11.2, includes the costs of ordering, purchasing, and maintaining these materials. At the beginning of each raw material purchase cycle, in which k will be equal to the production cycle, the seller spends O currency for the order and $C_R Q_R = C_R kmnq$ currency for the material purchase.

According to Figure 11.2, the current value of the cost of storing raw materials during a cycle (buying raw materials) is as follows:

$$rc_R \left\{ \sum_{j=0}^{k-1} e^{-z_j \frac{nq}{D}} \int_0^{\frac{nq}{R}} [mnq - mRt]e^{-zt}\, dt + \sum_{j=0}^{k-1} \int_0^{j\frac{nq}{D}} mnq e^{-zt}\, dt \right\} \tag{11.1}$$

Since the planning period is unlimited, the average value of the current annual cost of raw materials or the current value of raw materials cost during the planning period is obtained from the following sum.

$$RMC(q,n,k) = RMC_1(q,n,k) + e^{\frac{-zknq}{D}} RMC_1(q,n,k)$$
$$+ e^{\frac{-2zknq}{D}} RMC_1(q,n,k) + \ldots \tag{11.2}$$

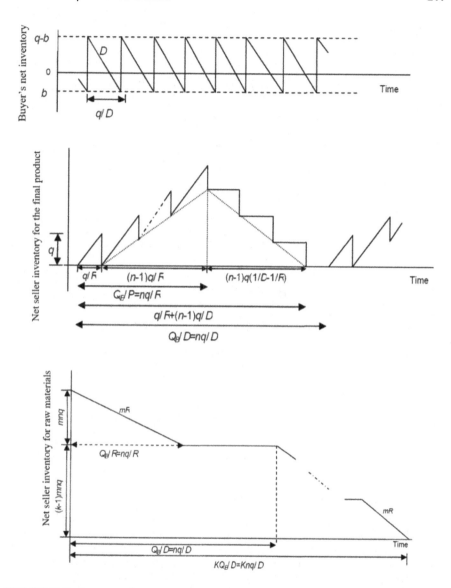

FIGURE 11.2   Inventory behavior: (a) buyer, (b) seller, and (c) net inventory.

The final product cost function for the seller

$$MFC(q,n) = MFC_1(q,n) + e^{\frac{-znq}{D}} MFC_1(q,n) + e^{\frac{-2znq}{D}} MFC_1(q,n) + \dots$$

$$= \frac{MFC_1(q,n)}{[1 - e^{-znq/D}]} \tag{11.3}$$

Buyer cost function

$$BC_1(q,b,n) = A + rc_b[\int_0^{\frac{q-b}{D}} (q-b-Dt)e^{-zt}dt]\sum_{j=0}^{n-1} e^{-zj\frac{q}{D}}$$

$$+ \pi[e^{\frac{-z(q-b)}{D}} \int_0^{\frac{b}{D}} Dte^{-zt}dt]\sum_{j=0}^{n-1} e^{-zj\frac{q}{D}}$$

$$+ (F+vq)\sum_{j=0}^{n-1} e^{-zj\frac{q}{D}} + c_b nq = A + rc_b[\frac{q-b}{z}$$

$$- \frac{D}{z^2}(1-e^{\frac{-z(q-b)}{D}})][\frac{1-e^{\frac{-znq}{D}}}{1-e^{\frac{-zq}{D}}}]$$

$$+ \pi e^{\frac{-z(q-b)}{D}}[\frac{D}{z^2}(1-e^{\frac{-zb}{D}}) - \frac{b}{z}e^{\frac{-zb}{D}}][\frac{1-e^{\frac{-znq}{D}}}{1-e^{\frac{-zq}{D}}}]$$

$$+ (F+vq)[\frac{1-e^{\frac{-znq}{D}}}{1-e^{\frac{-zq}{D}}}] + c_b nq \qquad (11.4)$$

Therefore, the current value of the buyer's costs will be:

$$BC(q,b,n) = BC_1(q,b,n)/[1-e^{-znq/D}] \qquad (11.5)$$

Integrated cost function

In integrated and coordinated chains, members make decisions that are optimized for the entire supply chain and minimize the integrated cost of the chain, $JTC(q,b,n,k)$. Although this decision may not be optimized for some members and may increase their costs, this increase will be compensated by dividing the total costs between the supply chain members and encouraging members to continue to cooperate.

$$JTC(q,b,n,k) = RMC(k) + MFC + BC(q,b,n) \qquad (11.6)$$

Solution algorithm

Metaheuristic algorithms can be used to solve nonlinear target functions with more than three decision variables that precise algorithms cannot be successful in solving for various reasons, such as the existence of integer variables and the complexity of the target function (Talbi 2009). In this section, two metaheuristic algorithms, that is, genetic and simulated annealing, are used. Also, these problems are solved by the optimization functions of MATLAB software to be compared with the answers of metaheuristic algorithms.

11.7 RESEARCH TRENDS

In the 1970s, inflation was considered an impact on inventory systems. Research shows that if the inflation rate is less than 4%, it can be ignored. If it exceeds 4%, however, it is necessary to include it in the decision, which will lead to an increase in the purchase amount. Demand is one of the parameters of the inventory system that may be discrete or continuous, fixed or variable. These are examples of different states of the demand parameter in an inventory system. Similarly, other common parameters in an inventory system, such as order costs, maintenance and shortage, commodity price, commodity life cycle, commodity corruption, supply time, receipt rate, number of warehouses, interest rate, and inflation rate can also be different. Each specific combination of these hypotheses may define a problem differently from the previous ones.

It is important to note that in the vast majority of research conducted in this area, the products in question have been perishable. Table 11.3 summarizes the research conducted in this area.

According to Table 11.2, studies show that unlike perishable goods, not much research has been done on obsolete goods; for this reason, it is interesting to study the issues that have been raised in the field of perishable goods and their implementation in the field of obsolete goods. One of the most important, and at the same time most practical, issues is considering inflation and its role in demand.

However, this section of the book attempts to cover some of the research gaps in the literature on integrated inventory models. But there is still a big gap between mathematical and theoretical models and the practical world. In this regard, the possibility of research in this field is wide, in which several suggestions for future research are presented. The development of co-economic accumulation determination issues in which demand depends on other parameters, such as quality, inventory level, time, inflation, and environmental factors, is proposed. Other dependency functions also seem useful and necessary. For example, in the inflation-related demand literature, different types of demand–inflation relationships are defined under different functions, some of which were mentioned in this chapter as an example.

11.8 A REVIEW: GENERAL DISCUSSION AND CONCLUSION

One of the most important issues in the world economy today is the phenomenon of inflation, which has wide economic, social, and political consequences. In general, it can be said that according to Dossche's theory, recognizing the dynamics of inflation is important in two ways (Dossche 2009). First, the existence of price-adjustment costs imposes additional costs on companies and changes the distribution of relative prices and demand. Second, it affects the real value of nominal assets, such as money.

TABLE 11.3

Classification of Research on Obsolete and Perishable Goods

Paper	Product type		Demand		Product price			Shortage		Supply chain type		# Of warehouses		# Of product levels		# Of chain levels			Type of chain components		Independent and dependent decision variables					
	Perishable	Obsolete	Uncertain	Certain	Price increase	Discount	Function of time, demand, the number, and combination of all three	backlog	Lost sale	Open	Closed	One	Two	One product	Multi-product	One	Two	Multi	EOQ	EPQ	Delay in payment	Inflation	Cost	Pricing	Optimal order time	Profit
Moussawi-Haidar et al. (2014)	*		*						*		*		*					*		*		*				
Yu, Mungan, and Sarker (2011)	*		*						*		*		*					*		*		*				
Sharma (2016)	*		*						*		*		*			*				*		*				
Zahran, Jaber, and Zanoni (2016)	*		*						*		*		*			*				*		*				
Zia and Taleizadeh (2015)	*		*				*		*		*		*							*		*				
Taleizadeh and Nematollahi (2014)	*		*				*	*	*	*										*	*	*				
Palanivel and Uthayakumar (2016)	*		*					*		*		*			*	*				*		*				
Yadav, Singh, and Kumari (2015)	*		*								*		*			*				*	*	*				
Tayal, Singh, and Sharma (2016)	*	*		*		*		*											*					*		
Nakandala, Lau, and Shum (2017)	*	*		*							*		*					*		*	*	*	*		*	*
Dai et al. (2018)	*	*		*	*			*				*		*					*		*			*	*	*

Paper — **Hypotheses and decision policies** / **Independent and dependent decision variables**

*Symbolizes that the relevant paper has that feature.

Category	Feature	Azadi et al. (2019)	(Shin et al. 2019)	Biuki, Kazemi, and Alinezhad (2020)	Yu et al. (2020)	Alkaabneh, Diabat, and Gao (2020)	Chen et al. (2020)
Product type	Perishable	*	*	*		*	*
	Obsolete	*	*	*		*	*
Demand	Uncertain						
	Certain	*		*	*	*	*
Product price	Price increase		*				
	Discount	*			*	*	*
	Function of time, demand, the number, and combination of all three	*	*	*	*		
Shortage	backlog	*	*	*	*		
	Lost sale			*		*	
Supply chain type	Open		*		*		
	Closed					*	
# Of warehouses	One			*			
	Two				*		
# Of product levels	One product		*	*	*	*	
	Multi-product						
# Of chain levels	One		*	*		*	
	Two			*			*
	Multi		*		*	*	
Type of chain components	EOQ				*		*
	EPQ		*	*	*		
Independent and dependent decision variables	Delay in payment				*	*	*
	Inflation		*				
	Cost			*	*	*	*
	Pricing		*	*		*	*
	Optimal order time					*	*
	Profit						

*Symbolizes that the relevant paper has that feature.

In an economy, changes in the distribution of relative prices and the real value of assets affect the allocation of community resources between consumption and investment. Inflation, on the one hand, imposes welfare costs by reducing the value of financial assets and has a kind of reducing effect on demand; on the other hand, by creating uncertainty in the decision of institutions to invest and creating other costs, it causes losses to production. Therefore, the implementation of optimal policies to deal with inflation requires an understanding of the dynamics of inflation. One of the most important and complex aspects of this dynamic is understanding the interaction between inflation and demand, which was the subject of this chapter of the book.

One of the basic characteristics of inventory-control systems and consequently developed models are various assumptions that exist in this field. Each of these parameters affect decision-making; depending on the industry under study and the conditions at the time of decision-making, the society may have different situations.

For example, demand is one of the parameters of the inventory system that may be continuous or discrete, fixed or variable, deterministic or probabilistic, dependent on other factors or independent of other factors, such as time, order amount, price of goods, commodity corruption, etc. The type of this dependence could take different forms such as linear, exponential, etc. These cases are an example of different states of demand parameters in an inventory system. Other common parameters in an inventory system include order costs, holding and shortage, commodity price, the period of life, corruption of goods, time of supply, commodity receipt rate, number of warehouses, interest rate, inflation rate, etc.

In the 1970s, due to the change in economic conditions, inflation was considered an effective factor in inventory systems. The results of different studies show that if the inflation rate is less than 4%, it can be overlooked. But in situations where this rate goes above 4%, considering it is necessary for decision-making, it leads to an increase in the purchase amount.

In this chapter, a sample model of inflation-dependent demand and, finally, the inventory system were investigated. The model tried to show a view of the effect of this important phenomenon on the inventory system. The results show that paying attention to the inflation phenomenon can be effective in applying the inventory model to real conditions.

REFERENCES

Alkaabneh, Faisal, Ali Diabat, and Huaizhu Oliver Gao. 2020. Benders decomposition for the inventory vehicle routing problem with perishable products and environmental costs. *Computers & Operations Research* 113: 104751.

Azadi, Zahra, Sandra D. Eksioglu, Burak Eksioglu, and Gokce Palak. 2019. Stochastic optimization models for joint pricing and inventory replenishment of perishable products. *Computers & Industrial Engineering* 127: 625–642.

Balkhi, Zaid T., and Lakdere Benkherouf. 2004. On an inventory model for deteriorating items with stock dependent and time-varying demand rates. *Computers & Operations Research* 31, no. 2: 223–240.

Ben-Daya, M., and S. A. Zamin. 2002. Joint economic lot sizing problem with stochastic demand. Systems Engineering Department, King Fahd University of Petroleum and Minerals, Dhahran, Saudi Arabia, *Technical Report.*

Biuki, Mehdi, Abolfazl Kazemi, and Alireza Alinezhad. 2020. An integrated location-routing-inventory model for sustainable design of a perishable products supply chain network. *Journal of Cleaner Production*: 120842.

Chen, Kebing, Tiaojun Xiao, Shengbin Wang, and Dong Lei. 2020. Inventory strategies for perishable products with two-period shelf-life and lost sales. *International Journal of Production Research*: 1–20.

Dai, Zhuo, Faisal Aqlan, Xiaoting Zheng, and Kuo Gao. 2018. A location-inventory supply chain network model using two heuristic algorithms for perishable products with fuzzy constraints. *Computers & Industrial Engineering* 119: 338–352.

Dossche, Maarten. 2009. Understanding inflation dynamics: Where do we stand? *National Bank of Belgium Working Paper*, no. 165.

Ghoreishi, M, A Mirzazadeh, and G.-W. Weber. 2014. Optimal pricing and ordering policy for non-instantaneous deteriorating items under inflation and customer returns. *Optimization* 63, no. 12: 1785–1804.

Gilding, Brian H. 2014. Inflation and the optimal inventory replenishment schedule within a finite planning horizon. *European Journal of Operational Research* 234, no. 3: 683–693.

Hou, K.-L., and L.-C. Lin. 2006. An EOQ model for deteriorating items with price-and stock-dependent selling rates under inflation and time value of money. *International Journal of Systems Science* 37, no. 15: 1131–1139.

Hsu, Ping-Hui, Hui Ming Wee, and Hui-Ming Teng. 2007. Optimal ordering decision for deteriorating items with expiration date and uncertain lead time. *Computers & Industrial Engineering* 52, no. 4: 448–458.

Maity, Kalipada, and Manoranjan Maiti. 2008. A numerical approach to a multi-objective optimal inventory control problem for deteriorating multi-items under fuzzy inflation and discounting. *Computers & Mathematics with Applications* 55, no. 8: 1794–1807.

Manna, S. K. 2019. An economic order quantity model for deteriorating item with non-linear demand under inflation, time discounting and a trade credit policy. *Pramana Research Journal* 9, no. 3,

Mirzazadeh, Abolfazl, M. M. Seyyed Esfahani, and S. M. T. Fatemi Ghomi. 2009. An inventory model under uncertain inflationary conditions, finite production rate and inflation-dependent demand rate for deteriorating items with shortages. *International Journal of Systems Science* 40, no. 1: 21–31.

Motallebi, S., and M. Zandieh. 2017. Determination of inventory management policies in process manufacturing: using discrete-event-simulation. *Journal of Industrial Management Perspective* 26: 83–108.

Moussawi-Haidar, Lama, Wassim Dbouk, Mohamad Y. Jaber, and Ibrahim H. Osman. 2014. Coordinating a three-level supply chain with delay in payments and a discounted interest rate. *Computers & Industrial Engineering* 69: 29–42.

Muniappan, P., R. Uthayakumar, and S. Ganesh. 2015. An EOQ model for deteriorating items with inflation and time value of money considering time-dependent deteriorating rate and delay payments. *Systems Science & Control Engineering* 3, no. 1: 427–434.

Nakandala, Dilupa, Henry Lau, and Paul K. C. Shum. 2017. A lateral transshipment model for perishable inventory management. *International Journal of Production Research* 55, no. 18: 5341–5354.

Pal, A. K., Asoke Kumar Bhunia, and R. N. Mukherjee. 2006. Optimal lot size model for deteriorating items with demand rate dependent on displayed stock level (DSL) and partial backordering. *European Journal of Operational Research* 175, no. 2: 977–991.

Palanivel, M., and R. Uthayakumar. 2016. Two-warehouse inventory model for non—instantaneous deteriorating items with optimal credit period and partial backlogging under inflation. *Journal of Control and Decision* 3, no. 2: 132–150.

Papachristos, Sotirios, and Konstantina Skouri. 2003. An inventory model with deteriorating items, quantity discount, pricing and time-dependent partial backlogging. *International Journal of Production Economics* 83, no. 3: 247–256.

Pareek, Sarla, and Vinti Dhaka. 2015. Fuzzy EOQ models for deterioration items under discounted cash flow approach when supplier credits are linked to order quantity. *International Journal of Logistics Systems and Management* 20, no. 1: 24–41.

Pathak, Savita, and Seema Sarkar. 2010. A three plant optimal production problem under variable inflation and demand with necessity constraint, imperfect quality and learning effects. *Journal of Computer and Mathematical Sciences* 1, no. 7: 769–924.

Pramanik, Prasenjit, and Manas Kumar Maiti. 2019. An inventory model for deteriorating items with inflation induced variable demand under two level partial trade credit: A hybrid ABC-GA approach. *Engineering Applications of Artificial Intelligence* 85: 194–207.

Sarker, Bhaba R., and Haixu Pan. 1994. Effects of inflation and the time value of money on order quantity and allowable shortage. *International Journal of Production Economics* 34, no. 1: 65–72.

Saxena, Seema, Vikramjeet Singh, Rajesh Kumar Gupta, Pushpinder Singh, and Nitin Kumar Mishra. 2020. A supply chain replenishment inflationary inventory model with trade credit. International Conference on Innovative Computing and Communications, University of Delhi, New Delhi, India.

Sharma, B. K. 2016. An EOQ model for retailers partial permissible delay in payment linked to order quantity with shortages. *Mathematics and Computers in Simulation* 125: 99–112.

Shin, Moonsoo, Hwaseop Lee, Kwangyeol Ryu, Yongju Cho, and Young-Jun Son. 2019. A two-phased perishable inventory model for production planning in a food industry. *Computers & Industrial Engineering* 133: 175–185.

Stadtler, Hartmut, Hartmut Stadtler, Christoph Kilger, Christoph Kilger, Herbert Meyr, and Herbert Meyr. 2015. *Supply Chain Management and Advanced Planning: Concepts, Models, Software, and Case Studies*. Heidelberg, Berlin: Springer.

Taheri-Tolgari, Javad, Abolfazl Mirzazadeh, and Fariborz Jolai. 2012. An inventory model for imperfect items under inflationary conditions with considering inspection errors. *Computers & Mathematics with Applications* 63, no. 6: 1007–1019.

Talbi, El-Ghazali. 2009. *Metaheuristics: From Design to Implementation*. Vol. 74. Hoboken, NJ: John Wiley & Sons Inc.

Taleizadeh, Ata Allah, and Mohammadreza Nematollahi. 2014. An inventory control problem for deteriorating items with back-ordering and financial considerations. *Applied Mathematical Modelling* 38, no. 1: 93–109.

Tayal, Shilpy, S. R. Singh, and Rajendra Sharma. 2016. An integrated production inventory model for perishable products with trade credit period and investment in preservation technology. *International Journal of Mathematics in Operational Research* 8, no. 2: 137–163.

Thangam, A., and R. Uthayakumar. 2010. An inventory model for deteriorating items with inflation induced demand and exponential partial backorders—a discounted cash flow approach. *International Journal of Management Science and Engineering Management* 5, no. 3: 170–174.

Tripathi, Rakesh. 2015. Economic order quantity for deteriorating items with non decreasing demand and shortages under inflation and time discounting. *International Journal of Engineering* 28, no. 9: 1295–1302.

Udayakumar, R., K. V. Geetha, and Shib Sankar Sana. 2020. Economic ordering policy for non-instantaneous deteriorating items with price and advertisement dependent demand and permissible delay in payment under inflation. *Mathematical Methods in the Applied Sciences*: 1–25.

Valliathal, M., and R. Uthayakumar. 2010. The production-inventory problem for ameliorating/deteriorating items with non-linear shortage cost under inflation and time discounting. *Applied Mathematical Sciences* 4, no. 6: 289–304.

Wee, Hui-Ming, and Sh-Tyan Law. 2001. Replenishment and pricing policy for deteriorating items taking into account the time-value of money. *International Journal of Production Economics* 71, no. 1–3: 213–220.

Yadav, Ajay Singh, and Anupam Swami. 2019. A volume flexible two-warehouse model with fluctuating demand and holding cost under inflation. *International Journal of Procurement Management* 12, no. 4: 441–456.

Yadav, Ajay Singh, Babita Tyagi, Sanjai Sharma, and Anupam Swami. 2017. Effect of inflation on a two-warehouse inventory model for deteriorating items with time varying demand and shortages. *International Journal of Procurement Management* 10, no. 6: 761–775.

Yadav, Dharmendra, S. R Singh, and Rachna Kumari. 2015. Retailer's optimal policy under inflation in fuzzy environment with trade credit. *International Journal of Systems Science* 46, no. 4: 754–762.

Yang, Shuai, Chulung Lee, and Anming Zhang. 2013. An inventory model for perishable products with stock-dependent demand and trade credit under inflation. *Mathematical Problems in Engineering* 2013.

Yu, Chunhai, Zebin Qu, Thomas W. Archibald, and Zhaoning Luan. 2020. An inventory model of a deteriorating product considering carbon emissions. *Computers & Industrial Engineering*: 106694.

Yu, Junfang, Deniz Mungan, and Bhaba R. Sarker. 2011. An integrated multi-stage supply chain inventory model under an infinite planning horizon and continuous price decrease. *Computers & Industrial Engineering* 61, no. 1: 118–130.

Zahran, Siraj K., Mohamad Y. Jaber, and Simone Zanoni. 2016. The consignment stock case for a vendor and a buyer with delay-in-payments. *Computers & Industrial Engineering* 98: 333–349.

Zia, Nadia Pourmohammad, and Ata Allah Taleizadeh. 2015. A lot-sizing model with backordering under hybrid linked-to-order multiple advance payments and delayed payment. *Transportation Research Part E: Logistics and Transportation Review* 82: 19–37.

12 Substitute and Complementary Goods

Shabnam Moradi
Department of Industrial Engineering, Amirkabir
University of Technology, Tehran, Iran

CONTENTS

12.1 INTRODUCTION

In today's competitive markets, considering the relationship between products can significantly affect firms' profits. Complementary and substitute products are the two main groups of products that play a crucial role in firms' strategies and customers' decisions. Understanding the effect and influence of each type of goods on demand function based on consumers' needs and wants is an obligation for each firm. In the following, these two types of goods are introduced.

12.1.1 SUBSTITUTE PRODUCTS

Companies or businesses are always facing some kind of competition. In most cases, substitute products are the primary source of this competition. In a market where there are fewer substitute products, it is possible to earn more profits. A substitute product is one that serves an equal purpose to another existing product in the market. Getting more of one substitute product lets the customer buy less of the other product. Thus, for any company, identifying substitute products that can perform the same function as its original product is an important issue.

Substitute products suggest some choices to consumers. Consumers' utility will increase if they make decisions by considering equally good alternatives. But, from a company's perspective, substitute products create competition. In other words, substitute products determine the intensity of competition and profitability. Sometimes, businesses are even put out of the race because substitute products considerably beat their offerings (Leo 1982).

Consumption of one product decreases or replaces the demand for the other substitutes; in other words, an increase in the price of one product will persuade customers to switch to other substitutes. Thus, the demand for substitute products displays a positive correlation with the product's price (Schwalbe and Zimmer 2009). For example, tea and coffee are substitutes for each other, and electricity is a substitute for solar energy. If the price of tea goes up, the demand for coffee goes up, too. It should be noted that this will only apply if we assume that the price of coffee remains stable.

12.1.2 COMPLEMENTARY PRODUCTS

Complementary products are those products that are not used independently but always used together to satisfy a specific need. Two products that complement each other reveal negative cross-elasticity. In other words, changing the price of a complementary product has an inverse relation with the demand for a given commodity. Thus, if the demand for a product increases, the demand for its complementary products will also increase and vice versa (Kumar and Sharma 1998). For example, if the price of pens increases, people will decide to use ballpoints. Consequently, the demand for ink will decrease significantly.

Complementary product pricing is a method used for stimulating the demand for other products. Complementary goods can increase profitability if the company uses them in the right way. As with all pricing methods, the company starts by learning how their customers make decisions. For complements, the company should be careful about which products or features customers consider when making their choices (Stiving 2011).

12.2 CONCEPTUAL FRAMEWORK

For analyzing demand rate and behavior based on price and inventory, it is necessary to study switching costs and various methods, such as bundling or package sales, besides looking at substitute and complementary types. Company managers will make more accurate decisions when they are informed regarding the mentioned topics. Their decisions concerning company product inventory level and price are more authentic to provide consumers the required demand level.

12.2.1 SWITCHING COST OF SUBSTITUTE PRODUCTS

The loss or extra cost that customers will afford if they decide to change a product to satisfy the same want is called switching cost. In many markets, there is a cost for consumers to change their supplier and switch to a competing firm. Consumers switching cost provides an opportunity for firms to exert market power over their permanent customers. This means the firm's future profit is highly dependent on its market share. In the case of low-priced products, such as stationery devices, there is a higher risk of consumers switching from a specific product to its substitute, except they have faith in a particular brand, which is called brand loyalty. Therefore, each company faces a trade-off between charging low prices and high prices for their products. Low prices provide the opportunity to attract new customers who would make repeated purchases, while high prices would lead firms to harvesting profits, which could run down the existing stock of market share (Klemperer 1995).

A variety of substitute products in the market decreases switching cost. The more substitute products are abundantly available in all corners of the market, the more consumers will switch their product (Klemperer 1995).

12.2.2 COMPLEMENTARY PRODUCT BUNDLING

Bundling is the idea of selling more than one item per package. Bundled packages usually include items that complement each other. In today's severe market atmosphere, companies must think through various matters, such as customer demand, product-specific costs, inventory level, and customers' multiple options. In addition, bundling decisions have substantial implications for monopoly power and marketing strategies. On the other hand, bundling can minimize customers' costs. These reduced costs depend on the number, value, and variation degree of items included in the package (Stremersch and Tellis 2002).

Complementary product bundling can suggest economies of scale.[1] Therefore, bundle choices and sizes are important for both customers and manufacturers. Furthermore, because of the negative cross-elasticity of demand for complementary products, the demand for one complementary good creates demand for the other. Companies can increase their marketing power by offering product packages, and retailers would be able to find efficient and viable prices. For example, if the main product is priced at a relatively low level compared to competitors, consumers would also consider the complementary product with a high probability (Yan and Bandyopadhyay 2011).

12.3 PRACTICAL EXAMPLE

The following case illustrates the price index numbers of complementary goods. In addition, it shows that practices and calculations in the realm of pricing without considering complementarity would cause incorrect results.

In this example, they estimate the price index that is obtained by considering the complementary relationship between inpatient days and physicians' operations. As a complement, price index differs substantially from the result of strategies without considering complementarity. First, the example shed light on why complementarity needs to be considered and how the implementation of complementarity in some cases can be a precise and simple solution to estimate quality changes. It was then concluded that the possible impacts of considering the complementarity of commodities on the pricing and inventory or volume are important for any kind of service. As a result, it is beneficial to perform experimental estimates in other fields, as the influence of complementary products is a necessary point for research in the pricing issue (Tuinen, Boo, and Rijn 1997).

Other instances close to this case are lumen hours of light (the combination of electricity, bulbs, etc.), vehicle kilometers (the variety of vehicles, fuel, repair services, road tax, insurance premiums, etc.), and medical treatment (combinations of bundles of medical services like inpatient care, operations by physicians, etc.). In other mentioned cases, the complexity for other techniques is higher due to the possible need for more data (Tuinen, Boo, and Rijn 1997).

12.4 DEMAND FUNCTIONS

Customer demand is influenced by many factors. Price and inventory level are two examples of these factors. The inventory of complementary goods seems to be more important because both goods are essential and should be bought together. Moreover, substitute products create competition. In addition, the cost of the substitute product should be less than the initial item. The circumstance may differ in the turnaround circumstance when the alternative's cost is more noteworthy than the first cost. The circumstance gets more complicated when the item falls apart in nature.

The linear model has been used widely in research (Choi 1996). This may be because not only is it easier and more understandable to solve problems and interpret them in this way, but it also helps to show the demand equations based on other equations, such as exponential functions, which make the calculations more complicated and require difficult justification. Table 12.1 categorizes the linear functions in which D_i: demand for product i, P_i: the price for product i, a_i: the demand base, and I_i: the stock level of product i.

The parameters are all non-negative in the first type of demand functions: β_{11} and β_{22} are larger than β_{12} and β_{21}; that is, the cross-price sensitivities are less than the self-price sensitivities. This shows that the demand for an item should be more dependent on fluctuations of its own price than on fluctuations in the price of its complement (Wei, Zhao, and Li 2013).

In the second type of demand function, to ensure that the total demand $D_1 + D_2$ is decreasing in P_1 and P_2, we shall assume that: $b_1 + \delta_1 - \delta_2 > 0$ and $b_2 + \delta_2 - \delta_1 > 0$ (Tang and Yin 2007).

12.5 EXAMPLES OF OPTIMIZATION MODELS

Here are some models that have been used in the literature.

12.5.1 SUBSTITUTABLE PRODUCTS MODEL

Krommyda, Skouri, and Konstantaras (2015) developed a stock-dependent demand model in which the products have a substitutable relationship.

The following notations are used to develop the model:

h_i The inventory holding cost for product i
p'_i The price of product i
c_i The unit purchase cost of product i, with $c_i < p'_i$
p_i The profit from selling a unit of product i ($p_i = p'_i - c_i$)

TABLE 12.1
Linear Demand Models

Type of function	Relation	Equation	Reference	Assumptions and notation
Price-dependent	Complementary	$D_{12}(P_1,P_2) = a - \beta_{11}P_1 - \beta_{12}P_2$ $D_{22}(P_1,P_2) = a - \beta_{22}P_2 - \beta_{21}P_1$	(Wei, Zhao, and Li 2013)	β_{ii}: the self-price sensitivities of a product's demand to its own price β_{ij}: the cross-price sensitivities $i \neq j$
	Substitute	$D_1 = a_1 - b_1 P_1 + \delta_1(P_2 - P_1)$ $D_2 = a_2 - b_2 P_2 + \delta_2(P_1 - P_2)$	(Tang and Yin 2007)	$a_i > 0$ δ_i: the notion of product substitutability a_i: the primary demand for product i
Stock-dependent	Complementary	$D_1 = a_1 + b_1 I_1(t) + b_3 I_2(t)$ $D_2 = a_2 + b_2 I_2(t) + b_3 I_1(t)$	(Hemmati et al. 2018)	b_1, b_2: sensitivity of each product's demand to its own stock level b_3: sensitivity of product's demand to the stock level of its complementary product $a_i > 0$ and $0 < b_i < 1$
	Substitute	$D_1(t) = a_1 + b_1 I_1(t) - b_2 I_2(t)$ $D_2(t) = a_2 - b_1 I_1(t) + b_2 I_2(t)$	(Krommyda, Skouri, and Konstantaras 2015)	$0 \leq t \leq T$, $a > 0$ and $0 < b < 1$
Bundling	Complementary	$D_3 = a - bP_3 + \lambda(P_1 + P_2 - P_3)$	(Yan and Bandyopadhyay 2011)	λ: the bundling discount price sensitivity P_3: the bundling price $0 < \lambda \leq b$ and $P_3 < P_1 + P_2$

A_i The ordering cost of product i
A The joint ordering cost for both products ($A<A_1+A_2$)
T The length of the replenishment cycle (a decision variable)
t_1 The time that the stock level of the first item reaches zero (a decision variable)
Q_i The order quantity of product i
x The fraction of demand of the product that has stocked-out that will be substituted by the other product during stock-out, $0 < x \le 1$
U The maximum stock capacity for both products
q The stock level of product 2 at time T (a decision variable)

The demand functions are:

$$D_1(t) = a_1 + b_1 I_1(t) - b_2 I_2(t) \tag{12.1}$$

$$D_2(t) = a_2 - b_1 I_1(t) + b_2 I_2(t) \tag{12.2}$$

where: $0 \le t \le T$ $a > 0$ and $0 < b < 1$.

The total profit per unit of time is:

$$\Pi(T,t_1,q) = \frac{1}{T}\left\{ P_1 Q_1 + P_2 Q_2 - h_1 \int_0^{t_1} I_1(t)dt - h_2 \int_0^{t_1} I_2(t)dt - h_2 \int_{t_1}^{T} I_2(t)dt - A \right\} \tag{12.3}$$

where:

$Q_1 = I_1(0), \quad Q_2 = I_2(0) - q.$

As a result, due to the existing limitations, the function will be maximized

$$p : \left\{ s.t. - \frac{(a_1+a_2)b_1}{b_1+b_2} \le \frac{f(T-t_1)}{(1-x)} e^{(b_1+b_2)t_1} \le \frac{(a_1+a_2)b_2}{b_1+b_2}, \atop 0 \le t_1 \le T, q \ge 0 \text{ and } I_1(0)+I_2(0) \le U. \right\} \tag{12.4}$$

with $\max_{(T,t_1,q)} \Pi(T,t_1,q)$

Unlike inventory and ordering cost, revenue has a direct relation with profit. By analyzing the feasible region of the previous equation, while Q only accepts a positive value, it can be solved by the derivative method (Krommyda, Skouri, and Konstantaras 2015).

12.5.2 BUNDLING

Taleizadeh et al. (2020) developed the following bundling model for complementary products:

λ Bundling discount price sensitivity
pi Selling prices of product i under a pure-selling strategy, $i=1,2$
T Common order cycle of products
T_B Common order cycle under the bundle policy
p_B Bundle price of products, $(p_B < p_1 + p_2)$
A_i Fixed ordering cost of product i
a The demand base
b Self-price sensitivity
h_i Holding cost of product i per unit time
h_B Holding cost of a bundle per unit time
c_i Unit purchasing cost of product i

$$D_B(p_1, p_2, p_B) = a - b p_B + \lambda(p_1 + p_2 - p_B) \tag{12.5}$$

The total profit of the retailer is:

$$\pi_B(T_B, p_B) = (p_B - c_1 - c_2)D_B(p_1, p_2, p_B) - \frac{A_1 + A_2}{T_B} + \frac{h_B D_B(p_1, p_2, p_B)T_B}{2} \tag{12.6}$$

where p_1 and p_2 are the selling prices of products 1 and 2 under the pure selling strategy.

Taking the first derivatives of Eq. (12.6) with respect to T_B and substituting $D_B(p_B)$, using Eq. (12.5) yields:

$$\frac{\partial \pi_B(T_B, p_B)}{\partial T_B} = \frac{A_1 + A_2}{T_B^2} - \frac{h_B(a - b p_B + \lambda(p_1 + p_2 - p_B))}{2} \tag{12.7}$$

On the other hand, taking the first derivatives of Eq. (12.6) with respect to p_B and setting it equal to zero yields:

$$\frac{\partial \pi_B(T_B, P_B)}{\partial p_B} = a - 2p_B(b + \lambda) + (C_1 + C_2)(b + \lambda) + \lambda(P_1 + P_2) + \frac{T_B h_B b + \lambda}{2} = 0 \tag{12.8}$$

The optimal bundle price, according to Eq. (12.8) is:

$$P_B^* = \frac{a + \lambda(P_1^* + P_2^*) + (\frac{T_B h_B}{2} + (c_1 + c_2))(b + \lambda)}{2(b + \lambda)} \tag{12.9}$$

12.6 RESEARCH TRENDS

Analyzing the demand rate requires studying the pricing policy and inventory level of each product of a company. Based on inventory and price, the products will be categorized into two main groups: complementary and substitute. Table 12.2 illustrates a summary of related research in this field. In Table 12.2, the reviewed papers and

TABLE 12.2

Research Trends of the Selling Price of Complementary Products and Substitutes[2]

Reference	Complementary	Substitute	SC structure	Objective function	Solution approach
(Taleizadeh et al. 2019)	*	*	One market, single retailer	Profit maximization	Classical optimization method
(Taleizadeh et al. 2020)	*		One market, single retailer	Price maximization	Hessian matrix
(Edalatpour and Mirzapour Al-e-Hashem 2019)	*	*	One supplier, one retailer	Profit maximization	Derivative method
(Lee 2017)	*		Two manufacturers, one common retailer	Inventory cost minimization	
(Wei, Zhao, and Li 2013)	*		One monopolistic retailer, two duopolistic manufacturers	Price maximization	Game theory
(Yue, Mukhopadhyay, and Zhu 2006)	*		Two suppliers, one market	Profit maximization	Bertrand type approach
(Giri, Mondal, and Maiti 2016)	*	*	Three manufacturers, one retailer	Profit maximization	Game theory
(Chen, Fang, and Wen 2013)		*	One manufacturer, an independent retailer, and an Internet channel	Profit maximization	Nash and Stackelberg games
(Hemmati, Fatemi Ghomi, and Sajadieh 2018)		*	Single vendor, single buyer	Profit maximization	Heuristic
(Jiang and Hao 2014)		*	Two manufacturers, uncertain markets	Profit maximization	Game theory
(Zhao et al. 2012)		*	Two manufacturers, common retailer	Profit maximization	Game theory

Reference	Complementary	Substitute	SC structure	Objective function	Solution approach
(Yan et al. 2014)	*		Single manufacturer, single market	Profit maximization	Derivative method
(Ren et al. 2020)	*		Two manufacturers, one retailer	Price maximization	Game theory
(Zhou et al. 2018)	*		One manufacturer, one retailer	Price maximization	Game theory
(Zhao, Wei, and Li 2014)	*		Two manufacturers, one retailer	Profit maximization	Game theory
(Mukhopadhyay, Yue, and Zhu 2011)	*		Duopolistic market, two firms	Profit maximization	Stackelberg
Hobara (2006)	*	*	Japan's free market	Utility maximization	Dynamic programming

works have concentrated on the operations management approach to confronting this chapter's problem. Yet, most related issues are modeled with linear demand function, which is not rational in the real world. As there is a significant expectation, the non-linear demand function is considered more often.

It is expected that companies and businesses rely on these studies more than before and use them in their real data. Implementing such models can be mutually beneficial for both consumers and firms. The substitution and complementarity effects cause a reduction in the firm's cost and enhancement in the quality of products. In this regard, translating this knowledge into practice has always been a common future challenge. Therefore, considering more realistic data and assumptions would be an excellent strategy to follow. For example, examining seasonal substitution and the pricing of complementary products and inventory control will illustrate this knowledge. Another example to consider is perishable products. If the exploration neglects the time parameter, the results will not be reliable.

It is acknowledged that there are still many outstanding issues, including the assessment of the change in the quality or utility of the given product or the influence of the appropriate pricing strategies on a wrong initial inventory decision. Further research is required to illuminate these areas.

NOTES

[1] Economy of scale: a proportionate saving in costs gained by an increased level of production.
[2] Most of the problems in this table are modeled with linear demand function, except the last one.

REFERENCES

Chen, Yun Chu, Shu Cherng Fang, and Ue Pyng Wen. 2013. Pricing policies for substitutable products in a supply chain with internet and traditional channels. *European Journal of Operational Research* 224, no. 3: 542–551.

Choi, S. Chan. 1996. Price competition in a duopoly common retailer channel. *Journal of Retailing* 72, no. 2: 117–134.

Edalatpour, M. A., and S. M. J. Mirzapour Al-e-Hashem. 2019. Simultaneous pricing and inventory decisions for substitute and complementary items with nonlinear holding cost. *Production Engineering* 13, no. 3–4: 305–315.

Giri, Raghu Nandan, Shyamal Kumar Mondal, and Manoranjan Maiti. 2016. Analysis of pricing decision for substitutable and complementary products with a common retailer. *Pacific Science Review A: Natural Science and Engineering* 18, no. 3: 190–202.

Hemmati, M., S. M. T. Fatemi Ghomi, and M. S. Sajadieh. 2018. Inventory of complementary products with stock-dependent demand under vendor-managed inventory with consignment policy. *Scientia Iranica* 25, no. 4: 2347–2360.

Hobara, Nobuhiro. 2006. The variety expanding growth model with change in substitution (complementary) among goods. *Hitotsubashi Journal of Economics*, no. 1: 265–283.

Jiang, Li, and Zhongyuan Hao. 2014. On the value of information sharing and cooperative price setting. *Operations Research Letters* 42, no. 6–7: 399–403.

Klemperer, Paul. 1995. Competition when consumers have switching costs: An overview with applications to industrial organization, macroeconomics, and international trade. *The Review of Economic Studies* 62, no. 4: 515–539.

Krommyda, I. P., K. Skouri, and I. Konstantaras. 2015. Optimal ordering quantities for substitutable products with stock-dependent demand. *Applied Mathematical Modelling* 39, no. 1: 147–164.

Kumar, Arun, and Sharma Rachana. 1998. *Managerial Economics*. India: Atlantic Publishers & Dist.

Lee, Young Hae. 2017. Optimum pricing strategy for complementary products with reservation price in a supply chain model. *Journal of Industrial & Management Optimization* 13, no. 3: 1553.

Leo, J. E. 1982. *Competitive Strategy: Techniques for Analysing Industries and Competitors: Porter.* New York: Michael E. Free Press/Macmillan.

Mukhopadhyay, Samar K., Xiaohang Yue, and Xiaowei Zhu. 2011. A Stackelberg model of pricing of complementary goods under information asymmetry. *International Journal of Production Economics* 134, no. 2: 424–433.

Ren, Minglun, Jiqiong Liu, Shuai Feng, and Aifeng Yang. 2020. Complementary product pricing and service cooperation strategy in a dual-channel supply chain. *Discrete Dynamics in Nature and Society*, no. 1.

Schwalbe, Ulrich, and Zimmer Daniel. 2009. *Law and Economics in European Merger Control.* Oxford: Oxford University Press.

Stiving, Mark. 2011. *Impact Pricing: Your Blueprint for Driving Profits.* New York: Entrepreneur Press.

Stremersch, Stefan, and Gerard J. Tellis. 2002. Strategic bundling of products and prices: A new synthesis for marketing. *Journal of Marketing* 66, no. 1: 55–72.

Taleizadeh, Ata Allah, Masoumeh Sadat Babaei, Shib Sankar Sana, and Biswajit Sarkar. 2019. Pricing decision within an inventory model for complementary and substitutable products. *Mathematics* 7, no. 7: 568.

Taleizadeh, Ata Allah, Masoumeh Sadat Babaei, Seyed Taghi Akhavan Niaki, and Mahsa Noori-Daryan. 2020. Bundle pricing and inventory decisions on complementary products. *Operational Research* 20, no. 2: 517–541.

Tang, Christopher S., and Rui Yin. 2007. Joint ordering and pricing strategies for managing substitutable products. *Production and Operations Management* 16, no. 1: 138–153.

Tuinen, Henk Van, Bram De Boo, and Jaco Van Rijn. 1997. *Price Index Numbers of Complementary Goods: A Novel Treatment of Quality Changes and New Goods, Experimentally Applied to Inpatient Medical Care.* Netherland: Statistics Netherlands.

Wei, Jie, Jing Zhao, and Yongjian Li. 2013. Pricing decisions for complementary products with firms' different market powers. *European Journal of Operational Research* 224, no. 3: 507–519.

Yan, Ruiliang, and Subir Bandyopadhyay. 2011. The profit benefits of bundle pricing of complementary products. *Journal of Retailing and Consumer Services* 18, no. 4: 355–361.

Yan, Ruiliang, Chris Myers, John Wang, and Sanjoy Ghose. 2014. Bundling products to success: The influence of complementarity and advertising. *Journal of Retailing and Consumer Services* 21, no. 1: 48–53.

Yue, Xiaohang, Samar K. Mukhopadhyay, and Xiaowei Zhu. 2006. A Bertrand model of pricing of complementary goods under information asymmetry. *Journal of Business Research* 59, no. 10–11: 1182–1192.

Zhao, Jing, Wansheng Tang, Ruiqing Zhao, and Jie Wei. 2012. Pricing decisions for substitutable products with a common retailer in fuzzy environments. *European Journal of Operational Research* 216, no. 2: 409–419.

Zhao, Jing, Jie Wei, and Yongjian Li. 2014. Pricing decisions for substitutable products in a two-echelon supply chain with firms different channel powers. *International Journal of Production Economics* 153: 243–252.

Zhou, Yancong, Jin Feng, Jie Wei, and Xiaochen Sun. 2018. Pricing decisions of a dual-channel supply chain considering supply disruption risk. *Discrete Dynamics in Nature and Society*, no. 1: 1–16.

13 Mass and Social Media

Mansoureh Naderipour
Department of Industrial Engineering, Amirkabir University
of Technology [Polytechnic of Tehran], Tehran, Iran

CONTENTS

13.1 INTRODUCTION

A network includes objects (nodes) and edges between nodes defining the connections between them. Social networks are a group of complex networks that create new problems and challenges in many fields, such as mathematics, biology, information science, sociology, and quantitative geography (Fortunato 2010). With the growth of social media like YouTube, Twitter, and Facebook, the nature of communication has been changed significantly from unidirectional to bidirectional, not only between firms and consumers but also among consumers. Generic Americans spend approximately 20% of their time on social media networks. From restaurant orders to electronics to travel to breakfast cereals, word of mouth and social media have become key sources of brand recommendations, loyalty, and new customer attainment (Hoffman and Fodor 2010).

Three types of search costs considering the incomplete information can be addressed (Stiglitz 1989). First, search costs of product information, especially the location of shops and prices, are the most commonly discussed search costs. When there is not enough information about the prices of goods in shops, or when the consumer does not know about the location of the (next) seller, or when getting the next store is too costly, it may lead to inefficient search or purchase. Next are searching costs for quality information, in particular for experiences, the quality of which becomes known when consumers have experienced them. As a result, consumers cannot derive the quality solely from the product or its services. Finally, identifying a product that fits consumers when products are incomplete substitutes may also have search costs for consumers. Most online market literature has discussed price and store search costs and shows how search and shopping engines affect consumer behavior and reduce consumer search costs (Brynjolfsson 2000, Bakos 1997). Actually, online markets occur between people who have never seen each other, thus sellers or buyers are especially "vulnerable to opportunistic behavior" (Ehavior and Pavlou 2002). Also, the uncertainty of quality could be higher for purchasers in online transactions compared to the traditional markets in which they may know the quality of product by "kicking the tires" (Chen, Wu, and Yoon 2004). One of the most important capabilities of the Internet is "bi-directionality," which results in sharing consumers' experiences and opinions easily compared to word of mouth in traditional transactions (Dellarocas 2003). Online word of mouth can reduce the uncertainty of a purchase quality and can be important information for customers (Chen, Wu, and Yoon 2004). One of the main questions is how online feedback mechanisms can increase confidence between different parts of online transactions and how variety ratings on marketers can make an impact on the number of bids, the probability of a sale, and prices. Nowadays, firms use "recommendation systems" and "customer profiling techniques" to reduce fit search costs to help consumers identify their desired products. As an example, Amazon.com uses a recommendation system so that it has a list of CD or book recommendations for customers by considering the information collected from other consumers (Dhar and Chang 2009).

Mass media is another source of information that plays an important role in people's lifestyles. Mass media refers to a diverse array of media technologies that reach a large audience via mass communication. The technologies through which this communication takes place include a variety of outlets. Mass media influences consumers' decisions through advertising and publicity (Tu et al. 2019, Ward 2001). Chapter 6 addresses the influence of advertising on demand; therefore, the aim of this chapter is to consider the impact of investments in technology on implicit communities and recommendation systems, especially to investigate the effectiveness of the factors of social networks, such as feedback and recommendation mechanisms, on consumer demand.

The organization of the chapter is summarized as follows: The next section presents the conceptual framework of the proposed factors. Then, the list of demand functions and their applications are presented. The relevant research trend is then reviewed, and finally, conclusions and suggestions for further research are given.

13.2 CONCEPTUAL FRAMEWORK OF THE PROPOSED FACTORS

In this section, the concepts related to the factors influencing the demand in social networks are given.

1. Consumer ratings: Consumers mostly pay attention to other consumers' reviews and experiences before buying products whose quality is uncertain. The feedback of the other consumers can be considered a quality index, which can reduce the uncertainty of product quality and can give them more confidence in buying the product (Chen, Wu, and Yoon 2004). As a result, when the uncertainty of the quality is reduced, it is observed that customers buy products with higher ratings.
2. Number of reviews: When the number of comments is high, it can be indicative of momentum of the product since consumers are mostly interested in exchanging reviews on products that are popular. More aspiring conversation may lead to more sales and customer interest, thus it is expected that a higher number of comments will result in higher sales and demand (Chen, Wu, and Yoon 2004).
3. Recommendation: The abundance of information is one of the most outstanding features of the Internet. Compared to traditional markets, in which consumers have to buy their products without enough information due to the difficulty of accessing information (i.e., information search cost is high), there are no such constraints in online transactions. Similarly, it is expected that recommendation systems can help consumers if they incur a high search cost for a desired product, and recommendations reflect the products that customers need (Chen, Wu, and Yoon 2004).
4. Activity: "Number of social network links (generated by comments/answers to users' posts)" (Chen, Wu, and Yoon 2004).
5. Group betweenness centrality: "Expresses the heterogeneity in betweenness centrality scores, which are a proxy of the brokerage power of users: they show how frequently a user is in-between the network paths that interconnect her/his peers" (Kadushin 2004).
6. Group degree centrality: "Measures how much variation there is in degree centrality scores of users, i.e. in their number of direct connections (the number of different people they interact with)" (Kadushin 2004).
7. Rotating leadership: "Sum of users' oscillations in betweenness centrality. A community where interaction is more 'democratic'—as members occupy less static positions—has more oscillations, which is usually beneficial to its participation and growth" (Newman 2004).
8. Sentiment: "Measures the positivity or negativity of the language used, with values in the range [0,1]. Neutral posts have a score of 0.5; higher scores indicate a more positive language" (Newman 2004).
9. Complexity: "Measures the complexity of the language used, with more complex posts having a higher score" (Newman 2004).

10. Average response time: "Average time taken by users to answer comments or questions (measured in hours)" (Newman 2004).

11. New users: "Counts the number of new users" (Newman 2004).

12. Google Trend: "Google Trend search volume index, for the search queries" (Newman 2004).

13. Word of mouth: Here, the focus is on the effect of word of mouth (WOM) on sales in a special category "A." Firms try to manage WOM between their customers. These firms talk to people about their products and try to increase the number of conversations. Therefore, firm-created WOM is a combination of traditional advertising and customer word of mouth so that the traditional advertising is "firm initiated" and "firm implemented," while the consumer word of mouth is "customer initiated" and "customer implemented" (Godes and Mayzlin 2009).

14. User-generated content and marketer-generated content: The impact of user-generated content (UGC) and marketer-generated content (MGC) on the valence of a product or brand in direct or undirected communications is an important issue. The valence obtained by UGC can be perceived as the customers' evaluations of a product or brand. Negative (or positive) valence of UGC can impede (or drive) customer purchases (Goh, Heng, and Lin 2013). The influence of MGC can be obtained from persuasive advertising, which includes messages that focus on and highlight a product's positivity to raise evaluations and promote good feelings among customers to encourage them to buy (Goh, Heng, and Lin 2013).

15. Communication intensity: Online social communications allow users to be aware of each other, and the awareness may increase with upgrading interactions. Different communication effects may come from different levels of consciousness (Goh, Heng, and Lin 2013).

16. Reviewer quality: Customers in online reviewing do not limit themselves to numeral scores, and they also pay attention to reviewer reliability. Online reviews may not be credible and trustworthy for customers. The customer writes reviews considering the quality of the product from the perspective of the user. The scores are mostly based on the customer's experience than on the characteristics of the product.

17. Reviewer exposure: Exposure and quality reputation are different. Here, exposure points to "media exposure of a reviewer in the online community." Exposure can be defined as "how many times a reviewer writes reviews" (Goh, Heng, and Lin 2013). Customers may attend more to higher exposure reviewers, in addition to higher quality, and they may not consider the reviews created by lower-exposure ones.

18. The role of product coverage and the age of an item: For products that have lower product coverage (with a smaller number of reviews), quality information is limited. Therefore, reviewers play an important role in informing customers about the quality of products and reducing uncertainty of these products. Each new reviewer may give additional information on quality to a new consumer. The effects of reviews on a product will be greater when there are fewer predefined reviewers. Therefore,

product coverage is the total number of customers reviewing a product (Hu, Liu, and Zhang 2008).

19. The difference between persuasive and informative advertising: The difference among "actual social learning channels," which have an effect on information about an unknown product, and "social persuasion channels," which can change the utility function of consumers, is considered here. In strong social learning, actual information about the production is shared, whereas in weak social learning, friends' product choices and their valuation for a new product are considered (Mobius, Niehaus, and Rosenblat 2005).

20. PageRank: PageRank measures the centrality of the product network position. The position of a product in the "copurchase network" is related to the network's impact on demand considering PageRank, a measure of centrality. "PageRank determines the ranking of the importance of web pages on the basis of the web created by the hyperlinks between the pages" by defining the following factors: first, "PageRank of the other product"; second, "The number of links originated from the other products" (Oestreicher-Singer and Sundararajan 2012).

13.3 LIST OF FUNCTIONS AND THEIR ADVANTAGES AND DISADVANTAGES

In this section, a list of functions based on social network factors affecting demand is given.

13.3.1 THE IMPACT OF ONLINE RECOMMENDATIONS AND CONSUMER FEEDBACK ON SALES BASED ON DATA GATHERED FROM AMAZON.COM

The following function is fitted considering the impacts of recommendations and consumer reviews. Table 13.1 summarizes the relevant terminologies:

Log form is used in this function as follows (Chen, Wu, and Yoon 2004):

$$\ln(rank_i) = \beta X_i + \gamma C_i + \lambda R_i + \varepsilon_i \tag{13.1}$$

X_i can be denoted as $X_i = [\ln(price_i), \frac{savings_i}{list\ price_i}, publish\ year_i, category_i]$. The impacts of mentioned variables are captured with β. Moreover, given that the level of demand for books is not the same, this function includes the subject of the book to consider the possible effect of book subject on demand ($Category_i$). The impacts of customer reviews on sales is captured with γ. The impact of recommendations on sales is indicated by λ. Here, lower ranks show higher sales; therefore, a "negative estimate indicates a positive relationship with sales and a positive estimate implies a negative relationship with sales" (Chen, Wu, and Yoon 2004).

People mostly exchange ideas that are "hot"; thus, more discussions may drive sales. This equation indicates that when sales data is not accessible, consumers may

TABLE 13.1

Summary of Equation (13.1) Symbols

Symbol	Description
$rank_i$	Sales rank of book i in Amazon.com
X_i	A set of control variables of book i
$price_i$	Price that is actually paid (i.e., list price savings)
C_i	The reviews of customers comprising the number of reviews for book i and the average score of customer reviews for book i
R_i	The number of recommendations of book i
ε_i	The random error

get reviews as a possible tool for sales. One of the problems with this discussion is that even though customer feedback information is available online, obtaining true sales and dealing data in online transactions is very hard. Also, customer ratings of books are not relevant to purchases because books are differentiated products that customers may choose without considering other consumers' ideas. Moreover, since most of books get high ratings, customers may not pay attention or consider them informative. In other words, customers may not take these ratings into account if they think the ratings are manipulated by others (Chen, Wu, and Yoon 2004).

13.3.2 THE EFFECT OF WORD OF MOUTH ON SALES: ONLINE BOOK REVIEWS

The sales rank of a book is a function of a "fixed effect" (υ_i), a "book-site fixed effect" (μ_i^A). Table 13.2 summarizes the relevant terminologies:

Therefore, the function is as follows (Chevalier and Mayzlin 2006):

$$\ln(rank_i^A) = \mu_i^A + \upsilon_i + \alpha^A \ln(P_i^A) + \gamma^A \ln(P_i^B) + X\Gamma^A + S\Pi^A + \varepsilon_i^A \qquad (13.2)$$

$$\ln(rank_i^B) = \mu_i^B + \upsilon_i + \alpha^B \ln(P_i^B) + \gamma^A \ln(P_i^B) + X\Gamma^B + S\Pi^B + \varepsilon_i^B \qquad (13.3)$$

"A" and "B" refer to Amazon.com and bn.com, respectively. There is a dummy variable that shows "usually ships in 24 hours," a dummy variable that shows "usually ships in 2–3 days," and other cases (Chevalier and Mayzlin 2006).

Then, if it is considered that $\mu_i^A = \mu_i^B$, Equation (13.4) is considered, and if $\mu_i^A \neq \mu_i^B$, Equation (13.5) is considered (Chevalier and Mayzlin 2006):

$$\ln(rank_i^A) - \ln(rank_i^B) = \beta^A \ln(P_i^A) + \beta^B \ln(P_i^B) + X\Gamma + S\Pi + \varepsilon_i \qquad (13.4)$$

$$\Delta[\ln(rank_i^A) - \ln(rank_i^B)] = \beta^A \Delta \ln(P_i^A) + \beta^B \Delta \ln(P_i^B) + \Delta X\Gamma + \Delta S\Pi + \varepsilon_i \qquad (13.5)$$

The advantage of Equation (13.4) is that we can apply more data since many reviews of books do not change during the time period. Moreover, estimating the coefficient

TABLE 13.2

Summary of Symbols of Equations (13.2) and (13.3)

Symbol	Description
υ_i	Fixed effect, such as quality of the book, popularity of the author, and offline promotion
μ_i^A	Book-site fixed effect: the fit between the book and the preferences of the customers of the site
P	Price
X	The vector of review variables of both sites (the impact of Amazon.com reviews on customers of bn.com and vice versa are considered)
S	A vector of dummy variables denoting the shipping times appointed by each website for each book

of price is allowed since there is not much difference in price over time. However, in Equation (13.5), a smaller sample should be used, and estimating all the interest coefficients is not allowed. However, it has the benefit of eliminating the "book-site-specific fixed" effect (Chevalier and Mayzlin 2006).

13.3.3 The Impact of User-Generated Content on Music Sales

The following function indicates the relationship between varieties factors and sales. In this equation, the usefulness of content comprising data from social network sites and blogs in determining sales in the music industry is considered (Dhar and Chang 2009). Table 13.3 summarizes the relevant terminologies:

$$\ln(Sales)_{i,t+n} = \alpha + \beta_1(\ln Blog\,Chatter_{i,t}) + \beta_2(NoMainstreamReviews_{i,t})$$
$$+ \beta_3(AvgConsumerRating i_{i,t}) + \beta_4(\ln\%ChgFriends_{i,t}) \qquad (13.6)$$
$$+ \beta_5(Chatter * Friends_{i,t}) + \beta_6(MajorIndieLabel_{i,t}) + \varepsilon_{i,t}$$

In this equation, $n = 1, 2, and\ 3$ means that sales are considered 1, 2, and 3 weeks in the future (Dhar and Chang 2009).

The distribution of blog chatter is considered log-normal, with "most albums receiving little attention but a few receiving a lot" (Dhar and Chang 2009). The level of album buzz is measured by the volume of received reviews, and quality is measured with the average positivity of the ratings. The review ratings are calculated by determining the number of review blogs that are considered numerical (for example, a 4-star scale). ChgFriends is considered as Myspace and shows social network intensity. There are four major groups in the music art; therefore, albums obtained by these four labels are more likely to have popular artists. For this reason, a dummy variable is defined for "major label effect," where 1 indicates a major label release and 0 indicates an independent label release (Dhar and Chang 2009).

TABLE 13.3

Summary of Equation (13.6) Symbols

Symbol	Description
i	The counter of album
t	Date
ε	An error term
BlogChatter	The total number of blog posts
ChgFriends	The number of friends an artist has on a social network

13.3.4 THE IMPACT OF SOCIAL NETWORKS ON ONLINE TRAVEL FORUMS AND TOURISM DEMAND

Determining the demand of tourism is an important issue for both policymakers and companies acting in the tourism industry. Fronzetti, Guardabascio, and Innarella (2019) considered social network to consider the content generated by users interacting on the TripAdvisor travel forum. The following function is proposed by Fronzetti, Guardabascio, and Innarella (2019). Table 13.4 summarizes the relevant terminologies:

$$y_t = \sum_{t=1}^{p} \phi_i y_{t-h-i} + \xi_j' F_{t-h} + \eta_t \quad t = 1,\ldots,T \tag{13.7}$$

ε_t is a "serially uncorrelated error term" with $E(\varepsilon_t) = 0$, $E(\varepsilon_t^2) = \sigma_\varepsilon^2$, $E(\varepsilon_t^4) < \infty$, so that $E(\varepsilon_t | y_{t-i})$. The h-step-ahead forecast is defined as follows:

$$\hat{y}_{t+h}^{FAAR} = \sum_{t=1}^{p} \hat{\phi}_i y_{T+1-p} + \hat{\xi}_j' \hat{F}_T \tag{13.8}$$

The factors are listed as follows:

- Percentage male: The proportion of male users who posted in the forum
- Average age: The average age of users who posted in the forum
- User levels: Each user activity on TripAdvisor is rewarded by a specific number of points. For example, 100 points for writing a review, 30 points for uploading a photo, and 20 points for writing a forum posts.
- User photos: The sum of the total number of photos uploaded on TripAdvisor by the users who were received in a city forum
- Activity: The number of social network links in a forum
- Group betweenness centrality: An indicator of how a user is in between the social network paths that interconnect users' peers
- Group degree centrality: The number of users they interact with
- Rotating leadership: The sum of users' oscillations in betweenness centrality

TABLE 13.4

Summary of Equation (13.7) Symbols

Symbol	Description
y_t	The target series
h	The number of steps ahead to forecast
ϕ_i	The ith coefficient of the autoregressive part of the model of order p
F_t	A vector of factors $(R \times 1)$
ξ	A coefficient vector

- Sentiment: The positivity or negativity of a language in the range [0,1]
- Complexity: The complexity of the language
- Average response time: Average time it takes users to answer comments or questions
- New users: The number of new users added to a forum
- Google Trend flights: The Google Trend search volume index for search queries made by the name of a city followed by the word "flights"
- Google Trend holidays: The Google Trend search volume index for search queries made by the name of a city followed (or preceded) by the word "holidays"

13.3.5 Firm-Created Word-of-Mouth Communication: Evidence from a Field Test

The following function studies the effect of word of mouth (WOM) communications on sales (Godes and Mayzlin 2009) Table 13.5 summarizes the relevant terminologies:

$$S_{it}^A = \sum_{j \in C,N} \left[\sum_{r \in R} \omega_j^r \cdot WOM_{ij(t-1)}^r \right] + \sum_{i=2}^{15} \mu_i + \sum_{t=2}^{12} \tau_t + \varepsilon_{it} \qquad (13.9)$$

Here, $r \in R = \{friend, relative, acquaintance, stranger, other\}$, and the focus is on the effect of WOM on sales of category A. μ_i terms include all systematic market-level factors, such as store location, competition, and market size. Focusing on these two factors, most of the seasonal and local shocks are considered.

13.3.6 Social Media Brand Community and Consumer Behavior: Quantifying the Relative Impact of User- and Marketer-Generated Content

The evaluation of a brand or product can be affected by the marketer and consumers. The following equations indicate the impact of user-generated content (UGC) and

TABLE 13.5

Summary of Equation (13.9) Symbols

Symbol	Description
S_{it}^A	Category A sales in market i in tth week
$WOM_{ij(t-1)}^A$	The set of potential relationships between the sender and receiver of WOM information
$WOM_{ij(t-1)}^r$	The number of reports in week t-1 in market i in condition j (i.e., customers or noncustomers) to consumers with whom they have communication, which can be denoted by r
C	Customer condition
N	Noncustomer condition
τ_t	Fixed effects of the week
μ_i	Fixed effects of the market

marketer-generated content (MGC) on the valence of a product or brand in direct or undirected communications (Goh, Heng, and Lin 2013). Table 13.6 summarizes the relevant terminologies:

$$UDVA_{it} = \frac{1}{J_{it}} \sum_{j=1}^{J_{it}} \left(\sum_{m=1}^{M_{ijt}} (UDVA_{ijtm} \times UIntensity_{ijtm}) \Big/ M_{ijt} \right) \qquad (13.10)$$

Dividing the weighted $UDVA_{ijtm}$ by M_{ijt} determines the "average valence of directed UGC from each customer j." For undirected communications, the following equation is defined (Goh, Heng, and Lin 2013):

$$UUVA_{it} = \sum_{n=1}^{N_{it}} UUVA_{itn} \Big/ N_{it} \qquad (13.11)$$

$$MDVA_{it} = \sum_{r=1}^{R_{it}} (MDVA_{its} \times MIntensity_{itr}) \Big/ R_{it} \qquad (13.12)$$

$$MUVA_{it} = \sum_{s=1}^{S_{it}} MUVA_{its} \Big/ S_{it} \qquad (13.13)$$

13.3.7 THE ROLE OF REVIEWER CHARACTERISTICS AND TEMPORAL EFFECTS

This research considers the effects of online reviews on sales and both qualitative and quantitative features of online reviews, such as product coverage, temporal effects, reviewer quality, and reviewer exposure. Customers read online comments and reviews and pay attention to other "contextual information," such as reviewer

TABLE 13.6

Summary of Symbols of Equations (13.10)–(13.13)

Symbol	Description
$UDVA_{ijtm}$	The valence of mth UGC, which customer i has observed from consumer j considering directed communication in period t
$UIntensity_{ijtm}$	The number of previous directed communications between customers i and j prior to mth directed communication in period t
M_{ijt}	The total number of UGC, which customer j has created for the customer i in period t by directed connection
J_{it}	The total number of customers who have created directed communications for customer i in period t
$UDVA_{it}$	The average valence of directed UGC for customer i
$UUVA_{it}$	The average v.alence of total N_{it} of UGC in period t, which customer i has observed from undirected messages and communications
$UUVA_{itn}$	The valence of the nth UGC in period t, which customer i has observed by undirected communication
$MDVA_{it}$	The average valence of directed MGC in period t, which the marketer has communicated to customer i for directed messages and communications
$MDVA_{itr}$	The valence of the rth directed MGC in period t, which the marketer has communicated to customer i
$MIntensity_{itr}$	The number of directed communications prior to rth directed communications between customer i and the marketer in period t
R_{it}	The total number of directed MGC in period t, which the marketer has communicated to customer i
$MUVA_{its}$	The valence of the sth MGC in period t, which customer i has observed by undirected communication
S_{it}	The total MGC in period t, which customer i has observed by undirected messages
$MUVA_{it}$	The average valence of MGC in period t, which customer i has observed by undirected message communication

exposure and reviewer reputation, in addition to the review scores. The market is more acceptable of reviews written by "reviewers with better reputation and higher exposure." The following function determines the impress of reviewer and product in sales (Hu, Liu, and Zhang 2008). Table 13.7 summarizes the relevant terminologies:

$$
\begin{aligned}
sales\_change = {} & \alpha_0 + \alpha_1 Signal + \alpha_2 Coverage + \alpha_3 Exposure \\
& + \alpha_4 Signal * Coverage + \alpha_5 Signal * Exposure \\
& + \alpha_6 * Book\_Dummy + \alpha_7 * DVD\_Dummy + \varepsilon_t
\end{aligned}
\tag{13.15}
$$

Signal is considered by the mentioned values: +1 indicates "favorable news by a high quality reviewer, 0 is a review by a low-quality reviewer," and −1 indicates

TABLE 13.7

Summary of Equation (13.15) Symbols

Symbol	Description
Signal	The level of quality of the reviewer and the types of reviewer (favorable or unfavorable)
Exposure	The influence of the reviewers' exposure on Amazon.com
Coverage	The level of reviewer coverage

"unfavorable news by a high quality reviewer." Signal shows a "qualitative property of the signal," which is sent to the Amazon.com market. Exposure equals 1 if "a review is written by a reviewer with more than the median number of exposures" and is otherwise 0. A dummy variable called "coverage" is defined and is equal to 1 if an item is followed by more than the median number of reviews, and 0 otherwise. Interactions between variables are considered, too: $Signal \times Exposure$, and $Signal \times Coverage$. Fixed effects considering item-level characteristics are indicated by product category dummies (Hu, Liu, and Zhang 2008).

13.3.8 SELF-SELECTION AND INFORMATION ROLE OF ONLINE PRODUCT REVIEWS

Here, the effectiveness of idiosyncratic preferences of early buyers on long-term customer buying behavior and "social welfare" constructed by review systems is discussed for online reviews of books on Amazon.com. The following function considers the structure of product ratings during the time (Li and Hitt 2008). Table 13.8 summarizes the relevant terminologies:

$$AvgRating_{it} = \begin{cases} \alpha + \beta \dfrac{T^{\lambda}-1}{\lambda} + u_i + e_{it} & \text{when } \lambda \neq 0 \\ \alpha + \beta.\log[T] + u_i + e_{it} & \text{when } \lambda = 0 \end{cases} \qquad (13.16)$$

$$\begin{aligned} -\log\,[salesRank_i] = \;& \beta_0 + \beta_1 AvgRating_i + \gamma_1 \log[P_i] \\ & + \gamma_2 Log[NumofReview_i] \\ & + \gamma_3 Log[P_i^C] + \gamma_4 Promotion_i \\ & + \gamma_5 T + \gamma_6 CatogoryDummies_i \\ & + \gamma_7 ShippingDummies_i + \varepsilon_i \end{aligned} \qquad (13.17)$$

To consider the influence of customer ratings on book sales, the factors related to demand are given (Li and Hitt 2008).

13.3.9 THE IMPACT OF SOCIAL MEDIA ON CONSUMER DEMAND: THE CASE OF THE CARBONATED SOFT DRINK MARKET

Choosing a particular brand from among competing brands and products using social media exposure and characteristics of products is discussed here. For these

TABLE 13.8
Summary of Equation (13.16) Symbols

Symbol	Description
$AvgRating_{it}$	The average review for book in time t
T	The time duration between when the average review was posted and the date the book was released
u_i	An idiosyncratic characteristic of each book that is kept constant during the time
p_i	The book price suggested by Amazon
$Numof\,Review_i$	The number of reviews posted on Amazon.com
p_i^C	The outside competitive price
$promotion_i$	Book promotion
$(CategoryDummies_i)$	Shipping availability

TABLE 13.9
Summary of Equation (13.18) Symbols

Symbol	Description
U_{ijm}	The utility of consumer j to purchase brand or product j in market m
p_{jm}	The unit price of brand j in market m
x_j	A vector of observed characteristics of brand j
ξ_{jm}	Unobserved product characteristics
SM_{jm}^{brand}	The social network exposure that considers the total communications and conversations about brand j
SM_{jm}^{price}	All communications about prices of brand j on social networks
SM_{jm}^{char}	A vector considering total communications about characteristic factors

reasons, the utility of the customer to buy a certain brand or product in a special market is considered (Liu and Lopez 2013). Table 13.9 summarizes the relevant terminologies:

$$U_{ijm} = \alpha_i p_{jm} + \beta_i x_j + \gamma_i SM_{jm}^{brand} + \phi_{1i} SM_{jm}^{price}$$
$$\times p_{jm} + \phi_{2i} SM_m^{Char} \times x_j + \xi_{jm} + \varepsilon_{ijm} \qquad (13.18)$$

For example, in this research, this function is considered for the carbonated soft drink (CSD) market and takes $x_j = (sugar_j, caffeine_j, sodium_j)$ as the observed nutritional characteristics of the CSD brand and $SM_{jm}^{char} = SM_{jm}^{nutrition} = (SM_m^{sugar}, SM_m^{caffeine}, SM_m^{sodium})$ (Liu and Lopez 2013).

13.3.10 SOCIAL LEARNING AND CONSUMER DEMAND

Here, the difference between persuasive and informative advertising is considered by finding the differences among "actual social learning channels," which have an effect on consumer information about an unknown product, and "social persuasion channels," which can change the utility function of consumers (Mobius, Niehaus, and Rosenblat 2005). Table 13.10 summarizes the relevant terminologies:

Consider the vector of the possible features of a product $m = (m_1,...,m_k,...,m_K)$, each of which can be 0 or 1 for implemented or implemented, respectively. υ_i depends on consumer appreciation of the features. The total value is considered as follows (Mobius, Niehaus, and Rosenblat 2005):

$$\upsilon_i = b_i'.m + \beta(d_{ij})\upsilon_j \tag{13.19}$$

This equation considers social influence. The first expression indicates the individual preferences of customer i. The second expression determines a customer's utility based on the valuation of customer j, who owns the product former considering the social distance d_{ij}. Every customer knows his/her preferences, but only the product users know the set of feature m. However, there are two ways for customer i to learn about the feature's value.

1. Strong social learning: The probability that customers i and j communicate with each other is $c(d_{ij})$. The probability that customer i learns about the full set of features (m) is indicated by $p(b'.m_i,\upsilon_j)$. This probability depends on the preferences of both customer i and customer j since the "communication is endogenous" (Mobius, Niehaus, and Rosenblat 2005).
2. Weak social learning: In some cases, customers i and j communicate with each other and customer i does not learn about the product's feature, while customer i does learn customer j's value $\upsilon_j = b_j'.m$. The probability of this occurrence is $q(b_i'.m_i,\upsilon_j)$, which is endogenous. It is assumed that customer i knows the degree of preferences that are correlated with the preferences of customer j: $h(d_{ij})$. The "actual degree of correlation" is indicated by ρ_{ij}, and the following function is determined (Mobius, Niehaus, and Rosenblat 2005):

$$E(b_i'.m|\upsilon_j) = h(d_{ij})\rho_{ij}\upsilon_j \tag{13.20}$$

TABLE 13.10
Summary of Equation (13.19) Symbols

Symbol	Description
υ_i	The valuation υ_i of the product, which consumer i gives
$b_{i,k}$	A value that customer i gives to feature m_k

13.3.11 The Visible Hand of Social Networks in Electronic Markets

By increasing electronic interactions, several visible and electronic networks have emerged, which has led to the connecting of businesses, products, and customers. The emergence of these networks affects many choices and demands in electronic markets. The demand can be measured considering two groups: first, the set of neighbors of the product that have a link to the product. It indicates products of which one of the co-purchase hyperlinks originates. Second is a set of complementary products. These products are often co-purchased with the product, whether there is a hyperlink between them or not. Considering these networks, the following function is defined for demand (Oestreicher-Singer and Sundararajan 2008). Table 13.11 summarizes the relevant terminologies:

$$y_i = \alpha_0 + \alpha_1 \sum_{j \in S_n(i)} y_j + \alpha_2 \sum_{j \in S_c(i)} y_j + \frac{1}{\alpha_n(i)} \sum_{w=1}^{K} \alpha_{3,w} \sum_{j \in S_n(i)} x_j + \sum_{w=1}^{K} \alpha_{4,w} x_i + \varepsilon_i \quad (13.21)$$

If N is the set of products in the whole recommendation network, $n = |N|$ can be denoted as the number of products in the whole recommendation network $N \subset M$. $\hat{y}$ indicates the $m \times 1$ vector $(\hat{y}_i)$, which is the vector y by adding rows at the bottom for products that are not in the recommendation network and are in the set of complementary products in the recommendation network. G_n is the $n \times n$ interaction matrix for set s_n, which is determined by $G_c(i,j) = 1$ if $j \in s_c(i)$; otherwise $G_c(i,j) = 0$ in which $s_n(i) \subset s_c(i)$. X indicates the product characteristic matrix $n \times K$, in which K is the number of characteristics. x_{ij} indicates the value of characteristic j for product i.

13.3.12 Recommendation Networks and the Long Tail of Electronic Commerce

The PageRank of a product measures the centrality of the product's network position. The position of a product in the co-purchase network is related to the network

---

TABLE 13.11
Summary of Equation (13.21) Symbols

Symbol	Description
y_i	The demand for product i
M	The set of all complementary products in the recommendation network
$m = |M|$	The number of products in M $(m > n)$
$s_n(i)$	The set of neighbors of product i that have a link to product i
$s_c(i)$	A set of complementary products to product i
$\sigma_n(i)$	The size of $s_n(i)$
y	An $n \times 1$ vector denoting demand

impact on the demand of the book considering PageRank, a measure of centrality. "PageRank determines the ranking of the importance of web pages on the basis of the web created by the hyperlinks between the pages" by defining the following function (Oestreicher-Singer and Sundararajan 2012):

$$PageRank(i) = \frac{(1-\alpha)}{n} + \alpha \sum_{j \in G(i)} \frac{PageRank(j)}{OutDegree(j)} \tag{13.22}$$

In this equation, $j \in G(i)$ if there is a connection from node j to node i, which means product j is in the "network neighbor of product i." Out degree (j) is the number of connections from product j (Oestreicher-Singer and Sundararajan 2012).

13.3.13 SELLING LUXURY FASHION ONLINE WITH SOCIAL INFLUENCES CONSIDERATIONS

In the luxury market, customers are divided into two groups of fashion leader and fashion follower. These two groups affect each other and make social influences in the retailing industry and market. The retailer sells fashion products to these two groups. Luxury fashion followers can see the Facebook, Twitter, and blog accounts of luxury fashion leaders, and luxury fashion leaders can see followers' comments on their social network sites to learn about the followers' demands. The demand for leaders is denoted by D_L, and the one for followers is shown with D_F and $D = D_L + D_F$. In the following function, if D_L increases, then D_F increases, and if D_F increases, then D_L decreases. The demand of the luxury fashion follower and luxury fashion leader are as follows (Shen, Qian, and Choi 2016). Table 13.12 summarizes the relevant terminologies:

$$D_F = \alpha_F + \Delta\alpha_F - \beta_F p + \eta_F s_F + \theta_F D_L \tag{13.23}$$
$$D_L = \alpha_L + \Delta\alpha_L - \beta_L p + \eta_L s_L - \theta_L D_F \tag{13.24}$$

$\Delta\alpha_i$ may be changed by sudden events, and it can be positive or negative (Shen, Qian, and Choi 2016).

TABLE 13.12
Summary of Symbols of Equations (13.23) and (13.24)

Symbol	Description
α_i, β_i, and η_i	The nonnegative variables
θ_i	A variable between [0,1]
p	The retail price per unit
s_i	The online retailer service
$\Delta\alpha_i$	The market demand change in a specific customer group

13.4 RESEARCH TRENDS AND FUTURE RESEARCH DIRECTIONS

It is worth mentioning that, for some products, such as books, no relationship has been found between customer ratings and sales. There can be many reasons for this. Given that books are differentiated products, customer tastes can be different, and they may take and like a book irrespective of how much others liked or disliked it. Moreover, as most of the books have high ratings, customers may not pay attention to these ratings. If customers find that these ratings may have been manipulated by others, they may even discount these ratings altogether. However, some research finds a positive relationship between sales and average ratings in general. The impact of retailer recommendations can be one of the reasons for research differences, which are not considered in some of them. Further research may be needed to match the findings. The prior studies and reviews are mentioned in Table 13.13.

The number of reviews has a positive effect on sales. Since consumers are more inclined to share their opinions on hot topics, the number of reviews a product has may cause product momentum in the market so that more aspiring conversations may cause more sales. However, higher reviews may be the result of higher sales. This indicates that when sales data are not attainable, the number of reviews or discussions can be used as a proxy for sales. This can have an important effect on future studies because usually the consumer information is available online, whereas transaction data are hard to obtain.

This chapter investigates the influence of social networks and media on demand. Future studies can be extended by a broader or deeper level of data in former research, such as customer comments, viewpoints of some "anchor readers," or using data from different contexts. For example, looking at products that are vertically differentiated[1] can have a higher influence on sales considering the reduction of quality uncertainty for them. Open issues considering causality also exist; for example, studying whether higher reviews cause more sales, which causes more reviews, or if they amplify each other.

TABLE 13.13
Prior Studies and Reviews

Study	Method	Data	Factors
Chen, Wu, and Yoon (2004)	Regression	Amazon Books 2003	Consumer ratings vs. sales, Number of reviews vs. sales, Recommendations vs. sales
Chevalier and Mayzlin (2006)	Regression	Books 2003–2004	Patterns of behavior of the reviewer, customer reviews
Hu, Liu, and Zhang (2008)	Regression	Books, DVDs videos	Reviewer quality, reviewer exposure, product coverage and age of an item
Oestreicher-Singer and Sundararajan (2008)	Regression	250,000 books sold on Amazon.com	The set of neighbors of a product, the size of it, the set of complementary products, the product characteristic matrix

(Continued)

TABLE 13.13

Prior Studies and Reviews

Study	Method	Data	Factors
Vasant Dhar and Elaine A. Chang	Regression	108 albums of music on Amazon.com	Blog chatter, mainstream media, average number of reviews and average rating, social networking intensity: MySpace, major label vs. independent label releases
Fronzetti, Guardabacio, and Innarella (2019)	Regression	TripAdvisor travel forum: 7 major European capital cities	User photos, user level, percentage male, average age, activity, group betweenness centrality, group degree centrality, rotating leadership, sentiment, complexity, average response ime, new users, Google Trend flights, Google Trend holidays
Godes and Mayzlin (2009)	Regression	Variety sites, such as Smoking Gun Slate, Slashdot, PostSecret, Google News, Friendster	Word of mouth (WOM)
Goh, Heng, and Lin (2013)	PSM	UGC and MGC data from an apparel retailer's brand community on Facebook	Marketer-generated content
Li and Hitt (2008)	Regression	Online book reviews posted on Amazon.com	AvgRating, the time duration, the idiosyncratic characteristic, number of reviews, outside competitive price, book promotion and shipping availability
Liu and Lopez (2013)	GMM	Carbonated soft drink market	Unit price of a brand, the vector of observed characteristics, the unobserved product characteristics, social network exposure, all communications about prices, vector considering total communications about characteristic factors
Mobius, Niehaus, and Rosenblat (2005)	Regression	Undergraduates at a large private university on Facebook	The value a customer gives each feature, strong social learning: the probability that customer i and j communicate with each other, the probability that customer i learns the full set of features, weak social learning:

Study	Method	Data	Factors
			the probability of this occurrence, the degree of preferences that correlate with the preferences of customer j, actual degree of correlation
Oestreicher and Sundararajan (2012)	Regression	Books in over 200 distinct categories on Amazon.com	PageRank of the other product, the number of links originated from the other product
Shen, Qian, and Choi (2016)	Game	Luxury fashion retailing industry	Luxury fashion retailing industry

NOTE

[1] "When people agree on which good is better than which, within the same category, we say these products are vertically differentiated. For example, people generally agree that a Mercedes car is better than a Yugo car."

REFERENCES

Bakos, J Yannis. 1997. Reducing buyer search costs: Implications for electronic marketplaces. *Management Science* 43, no. 12: 1676–1692.

Brynjolfsson, Erik. 2000. Frictionless commerce? A comparison of internet and conventional retailers. *Management Science* 46, no. 4: 563–585.

Chen, Pei-yu, Shin-yi Wu, and Jungsun Yoon. 2004. The impact of online recommendations and consumer feedback on sales. 25th International Conference on Information Systems (ICIS), Association for Information Systems, AIS Electronic Library (AISeL), Washington, DC. 711–723.

Chevalier, Judith A., and Dina Mayzlin. 2006. The effect of word of mouth on sales: Online book reviews. *Journal of Marketing Research* 43, no. 3: 345–354.

Dellarocas, Chrysanthos. 2003. The digitization of word-of-mouth: Promise and challenges of online feedback mechanisms. *Management Science* 49, no. 10: 1275–1444.

Dhar, Vasant, and Elaine A Chang. 2009. Does chatter matter? The impact of user-generated content on music sales. *Journal of Interactive Marketing* 23, no. 4: 300–307.

Ehavior, B, and Paul A Pavlou. 2002. Evidence of the effect of trust building technology in electronic markets: Price premiums and b uyer. *Ba & Pavlou/Trust Building Technology in Electronic Markets* 26, no. 3: 243–268.

Fortunato, Santo. 2010. Community detection in graphs. *Physics Reports* 486, no. 3–5: 75–174.

Fronzetti, Andrea, Barbara Guardabascio, and Rosy Innarella. 2019. Using social network and semantic analysis to analyze online travel forums and forecast tourism demand. *Decision Support Systems* 123: 113075.

Godes, David, and Dina Mayzlin. 2009. Firm-created word-of-mouth communication: Evidence from a field test. *Marketing Science* 28, no. 4: 721–739.

Goh, Khim-yong, Cheng-suang Heng, and Zhijie Lin. 2013. Social media brand community and consumer behavior: Quantifying the relative impact of user- and marketer- generated content. *Information Systems Research* 24, no. 1: 88–107.

Hoffman, Donna L., and Marek Fodor. 2010. Can you measure the ROI of your social media marketing? *MIT Sloan Management Review* 52, no. 1: 41–49.

Hu, Nan, Ling Liu, and Jie Jennifer Zhang. 2008. Do online reviews affect product sales? The role of reviewer characteristics and temporal effects. *Information Technology and managemente* 9: 201–214.

Kadushin, Charles. 2004. *Understanding Social Network.* Oxford: Oxford University Press.

Li, Xinxin, and Lorin M. Hitt. 2008. Self-selection and information role of online product reviews. *Information Systems Research* 19, no. 4: 456–474.

Liu, Yizao, and Rigoberto A. Lopez. 2013. The impact of social media on consumer demand: The case of carbonated soft drink market. Agricultural and Applied Economics Association's 2013 AAEA & CAES Joint Annual Meeting Washington, DC, August 4–6, Washington, DC.

Mobius, Markus M., Paul Niehaus, and Tanya S. Rosenblat. 2005. Social learning and consumer demand. International ESA Meetings and 2005 SITE Conference, Harvard University, Cambridge, MA.

Newman, M. E. J. 2004. Detecting community structure in networks. *The European Physical Journal B* 38: 321–330.

Oestreicher-Singer, Gal, and Arun Sundararajan. 2008. The visible hand of social networks in electronic markets. *Electronic Commerce Research.* New York: New York University.

Oestreicher-Singer, Gal, and Arun Sundararajan. 2012. Recommendation networks and the long tail of electronic commerce. *Management Information Systems Quarterly* 36, no. 1: 65–83.

Shen, Bin, Rongrong Qian, and Tsan-ming Choi. 2016. Selling luxury fashion online with social influences considerations: Demand changes and supply chain coordination. *International Journal of Production Economics* 185: 89–99.

Stiglitz, Joseph E. 1989. Imperfect information in the product market. In *Handbook of Industrial Organization*, eds. R. Schmalensee and R. Willig, 769–844. New York: Elsevier-Science.

Tu, Meng, Bing Zhang, Jianhua Xu, and Fangwen Lu. 2019. Mass media, information and demand for environmental quality: Evidence from 'under the dome. *Journal of Development Economics* 143: 102402.

Ward, Ronald W. 2001. A fresh meat almost ideal demand system incorporating negative TV press and advertising impact. *Agricultural economics* 25, no. 2–3: 359–374.

14 Past and Future of Demand Forecasting Models

T. Ahmadi
Center for Marketing and Supply Chain Management, Nyenrode
Business University, Breukelen, The Netherlands; Department
of Industrial Engineering and Innovation Sciences, Eindhoven
University of Technology, Eindhoven, The Netherlands

S. Solaimani
Center for Marketing and Supply Chain Management,
Nyenrode Business University, Breukelen, The Netherlands

CONTENTS

14.1 INTRODUCTION: THE IMPORTANCE OF DEMAND INFORMATION

Economics- and marketing-oriented research recognize that demand management can be interpreted as a firm's ability to identify customer demand and balance it with the firm's capabilities (Croxton et al. 2002; Rexhausen, Pibernik, and Kaiser 2012). In today's modern business workplace, demand management is not a disconnected back-office task. It is an indispensable element of a process that aims to link corporate strategic planning to daily operational plans by which the firm can balance demand with supply (Grimson and Pyke 2007). This process that integrates information across the supply chain (i.e., from the upstream to the downstream) is known as the sales and operations planning (S&OP) function.

By simplifying the supply chain into supply and demand activities, the S&OP function acquires input data from both the supply side and the demand side, which enables streamlined and coordinated operations or production and sales planning. Grimson and Pyke (2007) suggest that the S&OP department needs a cross-functional team in which there are representatives from the supply side (e.g., sourcing, production, and logistics) and the demand side (e.g., sales, marketing, and customer relationship management). In today's highly competitive global marketplace, firms need a well-aligned S&OP function to maintain their market value and increase their market share. The effectiveness of the S&OP function depends on whether firms can influence customer purchasing behavior.

According to Zhang and Zhang (2018), the market is a mixture of strategic and non-strategic customers, and therefore, the overall purchasing behavior is complex. More specifically, it is argued that a purchase does not happen instantly; instead, it occurs through a cycle with five steps, namely, (1) being stimulated, (2) search for relevant information, (3) evaluation of products, (4) making a purchasing decision, and (5) having post-purchase feelings, each of which can influence customer purchasing decisions (Lilien, Kotler, and Moorthy 1992). In sum, it is safe to say that determining customer demand based on customer purchasing behavior is not a straightforward task. As a result, the body of knowledge on how customer demand can be formulated and how it can be incorporated into operations decisions is ever-increasing in both marketing and operations domains. Commonly, researchers and practitioners model customer demand based on some influential operational and marketing factors.

In the simplest configuration, a supply chain can be considered a series of interconnected business entities in which the suppliers are located at the most upstream and the customers are located at the most downstream of the chain (Gligor 2014). The customers can be end consumers (e.g., in business-to-consumer settings) or non–end consumers (e.g., in business-to-business settings). The demand created by the non–end consumers is usually called derived demand. For the sake of consistency, henceforth, the term "customer demand" is used to refer to both demand types, unless a specific type of demand needs to be emphasized. It should be noted that in every supply chain, only one entity in the chain receives the end consumer demand, and the rest of the entities in the chain receive the derived demand (Mentzer and Moon 2004).

The point along the supply chain where the end consumer demand information penetrates the supply chain is called the customer order decoupling point (CODP). Depending on where the CODP is located along the supply chain, four main production strategies, namely, engineer-to-order (ETO), make-to-order (MTO), assemble-to-order (ATO), and make-to-stock (MTS), are distinguished (Atan et al. 2017). In these production strategies, activities located upstream of the CODP are managed using forecast-driven approaches (i.e., high uncertainty), and activities that are situated downstream of the CODP are managed using customer order–driven approaches (i.e., low uncertainty). Each of these four production strategies represents a specific trade-off between the cost-effectiveness of the operations and responsiveness toward customers. The accuracy of customer-demand information is critical in evaluating this trade-off.

In other words, overestimating and underestimating customer demand result in either an excess or shortage of supply, respectively. The excess of supply results in building up inventory, which in turn imposes inventory holding costs, such as opportunity cost of capital tied up in inventory, costs of storage and material handling, labor, and insurance costs. Besides, excess supply can lead to inventory obsolescence costs in companies with short life cycle products (e.g., high-tech companies with technological products).

On the other hand, a shortage of supply results in drawing down inventory and can result in late deliveries and customer dissatisfaction. In this situation, the customer might wait for the product and the demand is backlogged, or the customer might buy the product from a competitor and the demand becomes a lost sale (Ahmadi 2019). Even though, in reality, 85% of supply shortage cases result in lost-sale demand, because of the tractability of the analyses, researchers have analyzed operations decisions under the backlogged demand cases extensively (Bijvank and Vis, 2011, as cited in Ahmadi, Mahootchi, and Ponnambalam 2018).

The remainder of this chapter is organized as follows. We first introduce and discuss the most well-known quantitative forecasting techniques for independent- and dependent-demand models. Then, we discuss the forecast accuracy metrics and the accuracy of the historical data. Next, we explore the role of artificial intelligence (AI) in enhancing forecast accuracy. Finally, the chapter is concluded and suggestions for future research are provided.

14.2 QUANTITATIVE DEMAND FORECASTING MODELS

Demand estimation is a central task in retail operations and revenue management. For simplicity, many demand models in practice assume independent demand for each product. Scholars in operations and marketing science literature have developed various techniques for forecasting customer demand. Most demand forecasts rely on time-series models of historical data. However, the accuracy of the historical data as well as the performance of the forecasting technique may result in demand forecasts negatively or positively biased (i.e., underestimation or overestimation). The biased demand estimations might harm the performance of all the supply chain partners. In what follows, we review the most well-known forecasting techniques for independent-demand models (i.e., time-series) and dependent-demand models (i.e., causal models).

14.2.1 TIME-SERIES DEMAND MODELS

Nowadays, business firms record their transactions with each customer and store those in customer transaction databases. A time series refers to a series of historical data points labeled in time order. Let D_t represent the observed demand in period t, then forecasts are made with no additional information on future demand based on the time series $\overrightarrow{D_t} = \{D_t, D_{t-1}, D_{t-2}, \ldots\}$. According to Kremer et al. (2011), a time series can be modeled as $D_t = \alpha_t + \varepsilon_t$, where $\alpha_t = \alpha_{t-1} + \eta_t$. Random terms $\varepsilon_t \sim N(0, \delta^2)$ and $\eta_t \sim N(0, \sigma^2)$ are two independent random variables that capture temporary (i.e., the noise surrounding the level that lasts for a single period) and permanent (i.e., the notion of change in the true level α_t that persists in subsequent periods) shocks to the time series, respectively. Note that for $\delta = 0$, the demand model generates pure "random walks", and for $\sigma = 0$, it creates stationary "white noise" processes.

In general, time series can capture three main patterns in data, namely, level, trend, and seasonality. Let $\overline{D}_t$ be the average demand in period t. Then, the time series contains a level when the average demand is steady over time (i.e., $\overline{D}_t = \alpha_l$, where α_l is level). It contains a linear or non-linear trend when the demand has an upward growth or downward decline over time (i.e., $\overline{D}_t = \alpha_l + \alpha_\tau t$ or $\overline{D}_t = \alpha_l - \alpha_\tau t$, where α_τ is slope). It contains seasonality when a specific cyclic demand pattern is repeated within a time horizon (i.e., the season is defined based on the length of the cycle, such as year, season, month, week, etc.). The seasonality pattern can be modeled in an additive or multiplicative form. The former can be formulated as $\overline{D}_t = (\alpha_l + \alpha_\tau t) + \alpha_s(t)$ and the latter as $\overline{D}_t = (\alpha_l + \alpha_\tau t)\alpha_s(t)$, where $\alpha_s(t)$ is the seasonal component (Thomopoulos 2015, pp. 16–17). Figure 14.1 illustrates a time series containing level, trend, seasonality, and random noise components, that represents the historical sales records over 50 time periods. Notice that seasonality is added to the time series in an additive fashion.

To have more successful time-series extrapolation, choosing a proper quantitative forecasting technique based on the characteristics of the time series is required. For instance, when the time series just contains level and noise components, different techniques should be chosen than in a situation in which the time series has a trend or seasonality component. We categorize the traditional quantitative forecasting techniques in terms of their suitability for forecasting time series. Let F_t denote the

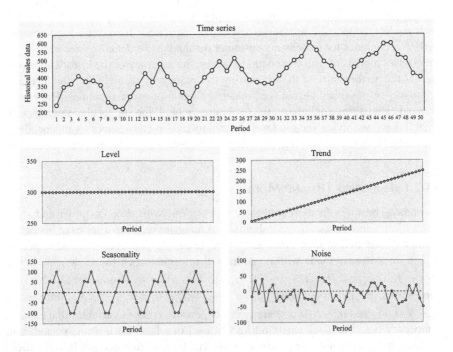

FIGURE 14.1   Time-series components.

demand forecast in period t, D_t denote the observed demand in period t, T be the length of time series, and h be the forecast horizon which is equal to the number of periods into the future for which forecasts are made.

14.2.1.1 Time-Series Forecast with Level

This category refers to a stationary time-series demand where the time series contains no long-run trend or seasonality components. For this category, the simplest forecasting technique is known as the naïve technique, in which the demand for the current period is considered as the forecast for the next period (i.e., $F_{t+1} = D_t$). However, this technique ignores all the old data; therefore, it is prone to statistical noise (Cachon and Terwiesch 2019, p. 270).

An approach that does incorporate all the old data into the forecast is simple average (SA), which is the arithmetic average of all the historical data. Hence, it can be formulated as

$$F_{t+1} = \frac{1}{t}\sum_{i=1}^{t} D_i.$$

One of the disadvantages of the SA technique is using all the historical data. It might be the case that the old data become harmful to the forecast over time. In other words, in the case of recent changes in the data, the SA technique lags behind the historical data, as it cannot eliminate the impact of old data on the forecast.

To overcome this issue, simple moving average (SMA_n) with an n-period time window is proposed in which just data from n recent periods are used in the forecast. Hence, we have

$$F_{t+1} = \frac{1}{n} \sum_{i=t-n+1}^{t} D_i.$$

As it is evident, the SMA_n gives the same level of importance to all the n recent data points, which are incorporated into the forecast, and ignores all the older data out of the time window. However, giving the same level of importance to all the recent n data points can be criticized since the older data should be valued less than the newer data. In other words, it is sensible that more importance is given to the most recent data, as they are a better candidate for forecasting the future. Following this line of reasoning, the weighted moving average ($WMA_{n,w}$) technique with an n-period time window and vector $\vec{w}$ is proposed as follows:

$$F_{t+1} = \sum_{i=t-n+1}^{t} w_i D_i.$$

where $\vec{w} = (w_{t-n+1}, w_{t-n+2}, \ldots, w_t)$ is a vector containing all the weighting parameters, such that $0 < w_{t-n+1} < w_{t-n+2} < \ldots < w_t \le 1$, and the summation of all the weights should be equal to 1. It means that different weights are given to the most recent n data points, such that data from the most recent period (i.e., $i = t$) gets the highest weight, and the data from the oldest period (i.e., $i = t - n + 1$) receives the lowest positive weight. In other words, $WMA_{n,w}$ gives zero weights to the oldest $t - n$ data points of the time series and gives positive and increasing weights to the last n data points of the time series. Yet, the total ignorance of the oldest $t - n$ data points, and also the distribution of the positive weights, might influence the forecast substantially.

As a solution to this issue, all the historical data are considered for the forecast, while weights decline exponentially with a fixed rate in terms of the age of the data. Based on this logic, Brown (1956) introduces the single exponential smoothing (SES_α) technique as

$$F_{t+1} = \alpha D_t + (1-\alpha)F_t$$

where α, $0 < \alpha < 1$ is called the smoothing level factor and F_0 should be initialized. Choosing the best value of the smoothing level factor depends on the characteristics of the time series. For instance, if the level of the time series changes slowly, then α should be chosen small to keep the impact of older data points. However, if the level of the time series changes quickly, then α should be large to reduce quickly the impact of older data points but not too large (Winters 1960). In other words, the bigger the value of α, the more SES_α is enabled to react to level changes quickly. Also, when the randomness of a time series is high, then α should be small to dampen the noise (Mentzer and Moon 2005, p. 88). Trigg and Leach (1967) propose adaptive α, in which the smoothing parameter for the next period is adjusted in terms of the forecast error of the current period. It is worth noting that SES_α with $F_1 = D_1$ is equivalent to $WMA_{n,w}$ with $n = t$ and $\vec{w} = ((1-\alpha)^{t-1}, \alpha(1-\alpha)^{t-2}, \alpha(1-\alpha)^{t-3}, \ldots, \alpha(1-\alpha), \alpha)$. In this setting, the summation of the weights is equal to 1; however, the increasing order of the magnitude of the weights is violated. For a more comprehensive comparison of the aforementioned forecasting techniques, see Chase (2013, p. 126).

14.2.1.2 Time-Series Forecast with Trend

When a product is introduced to the market for the first time or a competing product is introduced, a trend in their sales data will be marked (Winters 1960). When a time series contains a trend component, SES_α can no longer have a satisfying performance. As a more complete version of the SES_α, Holt's two-parameter approach, known as double exponential smoothing ($DES_{\alpha,\beta}$) with smoothing level factor α and smoothing trend factor β is introduced (Holt 1957). $DES_{\alpha,\beta}$ can be modeled as

$$L_t = \alpha D_t + (1-\alpha)(L_{t-1} + T_{t-1}),$$
$$T_t = \beta(L_t - L_{t-1}) + (1-\beta)T_{t-1},$$
$$F_{t+h} = L_t + hT_t,$$

where L_t is the forecasted level for period t, T_t is the forecasted trend for period t, α is the smoothing level factor, and β is the smoothing trend factor. Notice that L_0 and T_0 should be initialized, and the values of α and β should be chosen from the interval of (0,1) properly. Gardner and Mckenzie (1985) extend $DES_{\alpha,\beta}$

by adding an autoregressive-damping parameter (ϕ) to give more control over trend extrapolation as follows:

$$L_t = \alpha D_t + (1-\alpha)(L_{t-1} + \phi T_{t-1}),$$

$$T_t = \beta(L_t - L_{t-1}) + (1-\beta)\phi T_{t-1},$$

$$F_{t+h} = L_t + \sum_{i=1}^{h}\phi^i T_t.$$

According to Gardner and Mckenzie (1985), when $\phi = 0$, the proposed model is equivalent to the SES_α. When $0 < \phi < 1$, the forecasts approach an asymptote given by the horizontal straight line. When $\phi = 0$, the model is equivalent to the standard version of Holt's model. Finally, when $\phi > 1$, the trend is exponential.

14.2.1.3 Time-Series Forecast with Seasonality

This pattern can be seen in the sales data of seasonal products. For instance, the sales data of ice cream is repetitively high in the summertime and low in the wintertime. We expect that the same patterns will happen in each season of the upcoming years. To model the seasonality pattern, when the amplitude of the seasonal pattern is independent of the sales level, then an additive form (i.e., $D_t = \alpha_l + \alpha_\tau t + \alpha_s(t) + \varepsilon$), is recommended. However, when the amplitude of the seasonal pattern is proportional to the sales level, then a multiplicative form (i.e., $D_t = (\alpha_l + \alpha_\tau t)\alpha_s(t) + \varepsilon$), is recommended (Chatfield 1978; Winters 1960).

One of the approaches for forecasting a time series with seasonality is the decomposition technique based on the removal of the seasonal component. This may leave the sub-series without seasonality and with simpler patterns such as level or trend (Cleveland and Tiao 1976). In this procedure, first, the seasonal factor is calculated, and then using the seasonal factor a sub-series without the seasonality component is obtained (i.e., seasonally adjusted data). Then, a suitable forecasting technique is applied to the sub-series based on the characteristics of the sub-series. Next, the forecasts are re-seasonalized using the seasonal factor (Cachon and Terwiesch 2019, pp. 279–285).

Another approach that generalizes SES_α to cope with the trend and seasonality in the time series is introduced by Winters (1960) and is known as triple exponential smoothing ($TES_{\alpha,\beta,\gamma}$) with smoothing level factor α, smoothing trend factor β, and smoothing seasonal factor γ. In the literature, the $TES_{\alpha,\beta,\gamma}$ is also known as the Holt–Winters technique and is formulated as

$$L_t = \alpha\frac{D_t}{S_{t-s}} + (1-\alpha)(L_{t-1} + T_{t-1}),$$

$$T_t = \beta(L_t - L_{t-1}) + (1-\beta)T_{t-1},$$

$$S_t = \gamma\frac{D_t}{L_t} + (1-\gamma)S_{t-s},$$

$$F_{t+h} = \left(L_t + hT_t \right) S_{t-s+h},$$

where S_t estimates the seasonal factor for period t, and γ is the seasonal smoothing factor. We refer the reader to Chatfield (1978) for more detailed information on how the $\text{TES}_{\alpha,\beta,\gamma}$ technique can be applied to time series with seasonality in both additive and multiplicative forms.

14.2.1.4 Time-Series Forecast with Lumpy Data

In most business decision analyses, a continuous stream of positive time-series demands is considered. However, the demand may be lumpy, for which there are time intervals with no demand occurrences (i.e., intermittent demand) and large variations in the sizes of the demand when it is positive. Many service-oriented areas of the transportation/travel sector (e.g., airlines, hotels, cruise lines, and railways) face an intermittent or lumpy demand (Mukhopadhyay, Solis, and Gutierrez 2012). According to Chatfield and Hayya (2007), business decision analyses with lumpy demand become more challenging as less research exists to guide the decision-maker through the forecasting process.

Croston's approach (1974), in which analyzing the interval between consecutive non-zero demands and the volume of the non-zero demands is suggested separately, has been a base reference for forecasting lumpy time-series demand. Hence, for this type of demand, a forecast is made for when the next non-zero demand will appear and how large it will be. As an option, an all-zero forecast has been suggested when the demand is highly lumpy (Chatfield and Hayya 2007; Petropoulos et al. 2014). Syntetos and Boylan (2001) criticize the robustness of Croston's approach, and they identify several modeling limitations. They modify Croston's technique that gives approximately unbiased demand estimates per period. Chatfield and Hayya (2007) show that when the lumpiness of the time series is high, then the all-zero forecasting outperforms other forecasting techniques.

14.2.2 DEMAND CAUSAL MODELS

Demand causal models are formulated based on the causal relationship between demand and influential factors. The causal relationship can be modeled in various forms of linear, power, exponential, logarithmic, logit, and hybrid functions. In what follows, we review the regression and simulation-based models as common techniques to estimate parameters of the different demand functions.

14.2.2.1 Regression Models

Regression analysis is a statistical technique for estimating and forecasting the causal relationship between demand and other influential factors (e.g., product selling price, product freshness, product greenness) based on the historical data. The simplest regression model is the univariate linear regression model, which can be formulated as $D_i = a + bX_i + \varepsilon_i$, where D_i and X_i are dependent variable (i.e., the demand) and independent variable (i.e., the influential factor), respectively. Constant coefficients a and b are two parameters of the model that should be estimated based on historical paired data points (X_i, D_i) such that the sum of the squared errors is minimized.

By taking partial derivatives of the sum of the squared errors with respect to a and b and setting the derivatives equal to zero and solving the equations, the estimated values of $\hat{b}$ and $\hat{a}$ are estimated as

$$\hat{b} = \frac{\sum_{i=1}^{t} x_i d_i}{\sum_{i=1}^{t} x_i^2} \text{ and } \hat{a} = \overline{D} - \hat{b}\overline{X},$$

where $x_i = X_i - \overline{X}$ and $d_i = D_i - \overline{D}$ (Rawlings, Pantula, and Dickey 2001, pp. 3–4). Then, the demand forecast can be made using $F_i = \hat{a} - \hat{b}X_i$. A more complex regression model for forecasting demand can be considered by incorporating either more independent variables (i.e., multivariate regression) or a non-linear relationship between the demand and influential factors into the regression model (Tsekouras et al. 2007; Mohamed and Bodger 2005).

14.2.2.2 Simulation-Based Models

In this section, we review several simulation-based techniques for forecasting demand such as agent-based simulation (ABS), discrete-event simulation (DES), and system dynamics (SD). It is a widely held belief that SD is mostly used to model problems at a strategic level, whereas DES is used to model problems at an operational or tactical level (Tako and Robinson 2012). Using ABS models, the behavior of different influential agents (i.e., factors) in the system and their interactions with each other can be modeled. In other words, important macroscopic patterns from microscopic agents' behaviors can be achieved.

Simulation-based models are suitable since customers' purchasing decisions can be modeled using an ABS approach (Garcia 2005). Liang and Huang (2006) simulate a multi-agent supply chain and develop an agent-based demand forecast. Zhang and Levinson (2004) suggest ABS modeling as a travel demand forecasting technique that can predict important macroscopic travel patterns from microscopic agents' behaviors. Using the ABS models, different types of agents in the system and their interaction with each other can be incorporated into the customer demand forecast.

In SD models, it is essential to determine the influential factors in the system, their relationship with each other, and the dynamics in their behavior. Zhang, Zhang, and Zhang (2009) use SD simulation to consider population, economy, environment, and policy factors in forecasting water demand. Suryani, Chou, and Chen (2012) suggest that SD simulation is a suitable technique for forecasting air cargo demand where there exists system complexity with deep uncertainty. They find that SD not only can incorporate gross domestic product and foreign direct investment as two influential factors on cargo demand but can also incorporate expert knowledge into the simulation model with highly non-linear behavior. SD is also widely used for forecasting electricity demand in the energy market (He et al. 2017; Mirasgedis et al. 2006; Yan et al. 2018).

Mielczarek and Zabawa (2017) develop a hybrid simulation model based on DES and SD to investigate the influence of long-term population changes in healthcare services demand. They use an SD model to forecast the number of individuals who belong to their respective age–sex groups and a DES model to generate batches of patients with cardiac diseases and adjust the demand according to the demographic changes. Zhu, Hoon Hen, and Liang Teow (2012) use the DES model to estimate ICU bed capacity in a surgical ICU department of a Singapore government hospital. They simulate the complex patient flow with two sources of emergency and elective patients based on a first-come-first-serve discipline.

Although simulation-based models can be used in more complex settings (i.e., considering many factors and correlations in between), to provide insights into behavioral properties in a tractable way, they lack structural analysis. Nevertheless, simulation-based models are widely used for estimating demand in many domains, including the healthcare, energy, tourism and travel, airlines, food, and entertainment industries.

14.2.2.3 Artificial Intelligence Models

With the advent of advanced information technologies such as Internet of Things (IoT), cloud computing, artificial intelligence (AI), and Big Data infrastructure and analytical tools, a larger volume of data (i.e., real-time and near real-time) can be captured and analyzed, which helps to refine the existing forecasting models. Besides, user-generated (big) data is considered a useful source for monitoring and modeling people's intentions, preferences, and opinions (Brynjolfsson, Geva, and Reichman 2016). For instance, Schaer, Kourentzes, and Fildes (2019) investigate the usefulness of search traffic and social network shares in demand forecast over a product's life cycle, while Lau, Zhang, and Xu (2018) propose and evaluate a Big Data analytics method based on sentiment analysis to establish a better understanding of demand and to enhance sales forecasting.

Furthermore, AI techniques can improve the accuracy of the demand forecast substantially by incorporating more influential factors and modeling more complex relationships and interactions among factors using intelligent algorithms. Also, a hybrid version of AI and traditional forecasting techniques can be used. Rosienkiewicz (2020) proposes new hybrid models combining traditional forecasting techniques based on time series with AI-based methods for forecasting spare parts demand in the mining industry. The AI models need substantial amounts of data (Big Data) to outperform traditional forecasting models.

AI can also be utilized to estimate the parameters of the traditional forecasting techniques more intelligently (e.g., determining the best time window in the moving average techniques or smoothing factors in the exponential smoothing techniques). Moreover, AI can be used to evaluate the available traditional demand forecasting techniques and propose the most accurate one for a specific historical data set. Using machine-learning techniques (e.g., artificial neural networks [ANN], convolutional neural networks [CNN], Adaptive Boosting [AdaBoost], support vector machines [SVM], and random forests [RF]) and data mining techniques, complex forecasting models can be developed in which more factors (or dimensions of customer demand) can be incorporated into the formulation of the demand function. These data-driven methods can transform raw data into features spaces by finding very complex patterns in data of different sources (Wuest et al. 2016).

Up to date, different AI techniques for forecasting customer demand have been developed by scholars and practitioners for different applications. Grekousis and Liu (2019) use AI to improve demand predictability of emergency medical services. Huber and Stuckenschmidt (2020) employ AI to forecast customer demand in the retail industry, which requires daily forecasts. Sun et al. (2019) and Law et al. (2019) apply AI for forecasting demand in the tourism industry. Ryu, Noh, and Kim (2016) utilize AI for making a short-term forecast of electricity consumption. Güven and Şimşir (2020) use ANN and SVM for forecasting demand in the garment industry.

The machine-learning approaches provide a learning model for demand forecasting that can analyze influential factors, relationships, and complex interaction among factors from samples of the training data set. Machine learning needs substantial amounts of data (Big Data) and high processing power that is easily available nowadays. Huber and Stuckenschmidt (2020) argue that for applications with large-scale demand forecasting, machine-learning methods are more suitable forecasting techniques, as they provide more accurate forecasts. Gutierrez, Solis, and Mukhopadhyay (2008) use ANN modeling for forecasting lumpy demand. They compare the performance of the ANN forecasts with the traditional time-series forecasting techniques (e.g., the SES_α and Croston's technique) and find that the ANN model generally performs better than the traditional techniques based on three different performance metrics. In the context of short shelf-life food products, Doganis et al. (2006) use radial basis function neural network architecture for building the time-series model and a specially designed genetic algorithm to select the appropriate input variables to the model, leading to a more accurate approach with lower forecast error when compared to other time-series methods, such as linear autoregressive and Holt–Winters.

In sum, the power of AI models can be used for forecasting both independent- and dependent-demand models with any level of complexity. To do so, traditional forecasting techniques can be used for developing machine learning or deep-learning forecasting algorithms.

14.3 FORECAST ACCURACY METRICS

To evaluate the accuracy of the forecasting techniques, different forecasting error measures, such as mean absolute error (MAE), mean square error (MSE), and mean absolute percentage error (MAPE), have been developed (Hyndman and Koehler 2006). These metrics incorporate the forecast error, E_t, into their measurement. E_t is defined as the difference between the observed data and its forecast; $E_t = D_t - F_t$.

In the literature, MAE is also known as mean absolute deviation (MAD) and both MAE and MAD have been used interchangeably. The MAE is the most straightforward metric to calculate and to understand as it can be formulated as

$$MAE = \frac{1}{n}\sum_{t=1}^{t}|D_t - F_t|.$$

However, since it is a scale-dependent metric, it is not sensible to utilize this metric to compare the forecast accuracy across different time series with different scales.

The MSE metric is useful for comparing forecast accuracy under quadratic loss. In essence, it penalizes large forecast errors more severely than other common accuracy statistics (Thompson 1990). In other words, it exaggerates the magnitude of the forecast error by taking it to the power of two. Then, it can be formulated as

$$MSE = \frac{1}{n}\sum_{t=1}^{n}(D_t - F_t)^2.$$

MSE is also a popular metric due to its theoretical relevance in statistical modeling (Hyndman and Koehler 2006). Similar to the MAE metric, the MSE is a scale-dependent metric, and therefore it is not applicable to compare forecast accuracy across different time series with different scales.

MAPE is a perfect metric for comparing forecast accuracy across time series with different scales (i.e., scale-independent metric), and it is modeled as

$$MAPE = \left(\frac{1}{n}\sum_{t=1}^{n}\frac{|D_t - F_t|}{D_t}\right)\times 100\%.$$

However, MAPE has the disadvantage of being infinite or undefined when the time series contains zero values (Hyndman and Koehler 2006). For instance, this metric cannot be applied to lumpy or intermittent time-series demand that contains zeros.

It is worth mentioning that the MAE and MSE metrics, whose scales depend on the scale of the data, are useful for comparing the performance of different forecasting techniques applied to the time series with the same scale. However, they should not be used for comparing across time series that have different scales. To employ the MSE metric for assessing overall accuracy across many series, Thompson (1990) proposes a modified version of MSE statistic as log mean squared error ratio (LMR).

14.4 DATA ACCURACY

Typically, historical data is based on the sales transactions in which spilled (i.e., demand lost due to unavailability of the consumer's first choice) and recaptured (i.e., demand for a substitutable product due to unavailability of the customer's preferred product) demand are not readily observable (Vulcano, van Ryzin, and Ratliff 2012). Nevertheless, different statistical techniques, known as demand untruncation or demand uncensoring methods, are proposed to estimate spilled demand and recaptured demand. One of the most applied methods is the expectation-maximization algorithm that employs iterative methods to estimate the underlying parameters of interest (Vulcano, van Ryzin, and Ratliff 2012). Another widely used approach in determining the demand for various products within a set of comparable items is to utilize discrete choice models, which predict the likelihood of customers purchasing a specific item from a set of related items based on their relative attractiveness (Vulcano, van Ryzin, and Ratliff 2012).

In the era of globalization, supply chain management becomes more and more challenging as the complexity of the supply chain networks (i.e., the number of echelons, the number of entities at each echelon, and their connectivity within and across the echelons) increases and the supply chain visibility decreases. As a result, customer demand information is distorted at the upstream supply chain entities. The distortion amplifies the further away suppliers are situated from the focal company; a phenomenon known as the Bullwhip effect, observed and discussed by Forrester (1958). One crucial mechanism for coordinating the supply chain is information flow and knowledge sharing among the supply chain entities (Solaimani et al. 2015).

To study the impact of supply chain coordination, the following form of the stochastic demand process has been commonly used (Chen et al. 2000). Let $\mathcal{D}_t$ represent the stochastic demand variable in period t, then $\mathcal{D}_t = \alpha_0 + \rho \mathcal{D}_{t-1} + \varepsilon_t$, where α_0 is a nonnegative constant, ρ is a correlation parameter with $|\rho| < 1$, and the error terms, ε_t, are independent and identically distributed (IID) from a symmetric distribution with zero mean and standard deviation σ. It can easily be shown that $\mathbb{E}\{\mathcal{D}_t\} = \dfrac{\alpha_0}{1-\rho}$ and $\mathrm{Var}\{\mathcal{D}_t\} = \dfrac{\sigma^2}{1-\rho^2}$. Notice that when $\rho = 0$, the demand over periods is IID with mean μ and standard deviation σ.

There is a consensus among scholars and practitioners that obtaining and sharing customer demand information with all the partners of a supply chain enhances coordination and the system's performance. The role of sharing customer demand information in supply chain performance is emphasized when this information is shared in advance, which is called advance demand information (ADI). With the help of advances in information technology, such as electronic data interchange (EDI), web-based platforms, and Internet-based communication tools, business firms can obtain ADI from their customers by implementing preorder strategy, in which the customers place their order ahead of time to share the information on timing and sizes of their future order with the firm. Sharing ADI between customers and sellers creates a win-win situation for both parties based on which excess inventory and shortage of supply (or product unavailability) can be mitigated. ADI can also reduce customer waiting time and increase the service level (Ahmadi, Atan, Kok, and Adan 2020).

Suppose customers need a product at time t_1. Under ADI, customers place their orders at t_0 with the firm, where $t_0 < t_1$ (i.e., $D(t_1) = O(t_0)$). In other words, the customer order at the time t_0 turns to customer demand at the time t_1. If the firm cannot fulfill the demand at the time t_1, then the customer needs to wait until the product is available. As illustrated in Figure 14.2, the time window between t_0 and t_1 is called commitment lead time.

Based on the accuracy of the information, ADI can be categorized as perfect ADI (i.e., customers share exact information on the timing and size of their future orders with the supplier) or imperfect ADI (i.e., customers provide the firm with an estimate of timing or quantity of future orders, which may then later be modified or canceled). For detailed information on perfect and imperfect ADI, see Ahmadi, Atan, Kok, and Adan (2019a,b), and the references therein.

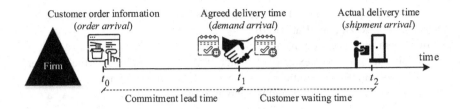

FIGURE 14.2   The occurrence of customer order and customer demand with ADI.

14.5 CONCLUSION AND FUTURE RESEARCH DIRECTIONS

In this chapter, we discussed how demand management plays a substantial role in balancing demand with supply in modern business workplaces. This process integrates information across the supply chain known as the sales & operations planning (S&OP) function, which is comprised of a cross-functional team with representatives from the supply side and the demand side. The S&OP function acquires input data from both the supply and demand sides. The input from the demand side requires estimation models that can predict customer purchasing behavior.

We categorized the quantitative forecasting methods based on demand dependency as independent- and dependent-demand models. For the independent models, we reviewed time-series forecast models in which the dependent demand is forecasted based on the historical sales records in the form of time series. For the dependent models, we reviewed techniques for estimating the parameters of the causal models, in which the dependent demand is modeled in terms of some influential factors (e.g., product selling price, customer willingness-to-pay, product greenness, product freshness, product visibility, refund policy, and service quality) that can influence the customer purchasing decision. As illustrated in Figure 14.3, for each demand type, various forecasting methods have been reviewed and discussed.

For the time-series demand models, we noticed that identifying the patterns of the time series (i.e., level, trend, and seasonality) facilitates the selection of a forecasting technique. Depending on the characteristics of the time series, we classified the traditional time-series forecasting techniques. For stationary time series without trend and seasonality patterns (i.e., with level and random noise), we introduced naïve, simple average (SA), simple moving average (SMA_n), weighted moving average ($WMA_{n,w}$) with an n-period time window and weighting vector w, and single exponential smoothing (SES_α) with smoothing level factor α. We discussed the pros and cons of each method.

For a time series without seasonality and with a trend, we reviewed Holt's technique, known as double exponential smoothing ($DES_{\alpha,\beta}$ with smoothing level factor α and smoothing trend factor β. We also discussed the damped version of the double exponential smoothing that adds a damping parameter to $DES_{\alpha,\beta}$ to give more control over trend extrapolation. For a time series with seasonality, patterns can be added to the time series in either an additive or multiplicative form. For a time series with seasonality, we introduced the decomposition technique, in which the first data series is seasonally adjusted, then the seasonally adjusted data is forecasted using a

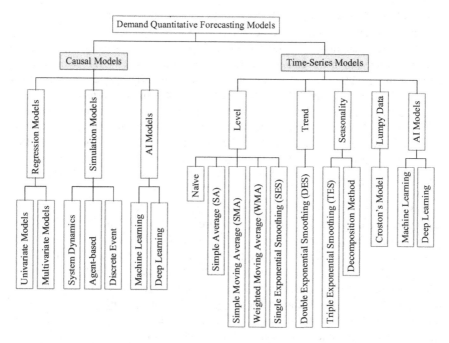

FIGURE 14.3 Demand quantitative forecasting techniques.

proper forecasting technique. Finally, the forecasts are reseasonalized. Furthermore, we discussed the Holt–Winters technique known as triple exponential smoothing ($\text{TES}_{\alpha,\beta,\gamma}$) with smoothing level factor α, smoothing trend factor β, and smoothing seasonal factor γ.

For the causal models, univariate and multivariate regression models and simulation-based models were discussed. We reviewed how simulation-based models in the form of discrete-event, system dynamics, and agent-based simulation models applied to different applications.

We also reviewed scale-dependent forecast accuracy metrics, such as mean absolute error (MAE), mean square error (MSE), and scale-independent forecast accuracy metric mean absolute percentage error (MAPE). We discussed their characteristics, advantages, and disadvantages. The scale-dependent metrics should not be used for comparing the forecast accuracy across different time series with different scales, while scale-independent metrics can be applied. In general, the accuracy of a forecast depends on the amount of available historical data, data features (e.g., patterns and amount of randomness), forecast horizon, forecasting technique, and forecast accuracy metric (Petropoulos et al. 2014). Another element that impacts forecast accuracy is the accuracy of the historical sales data, which is used as input in the forecast. We discussed that spilled and recaptured demand are not readily observable. We also discussed advance demand information (ADI) in both perfect and imperfect forms. In this setting, customers share their demand information (i.e., information on timing and the size of the order) in advance of their actual needs.

Even in imperfect form, in which customers can cancel or change their order information, ADI can boost forecast accuracy.

Besides all the traditional forecasting techniques for time-series data and estimating parameters of the causal models, artificial intelligence (AI) techniques can improve the accuracy of the demand forecast substantially by incorporating more influential factors and modeling more complex relationships and interactions among factors using intelligent algorithms. We reviewed how machine learning techniques have been applied to forecast demand for different applications. Even though the advanced forecasting technics based on AI can outperform traditional methods, they also have some drawbacks. The AI models do not provide analytical insights and require a large amount of data and processing power, especially in deep-learning models.

Although businesses usually have access to advanced and sophisticated quantitative methods embedded in forecasting software, empirical evidence reveals that real-world forecasting relies typically on human judgment. In other words, as the first step, computer-based quantitative methods provide the basis for a forecast, and it is modified based on human judgment in the next step (Kremer, Moritz, and Siemsen 2011). No forecasting technique fits all types of data series. The most suitable forecasting technique for each dataset should be found among possible candidates based on the characteristics of the data series. In this setting, AI can be used not only for the forecasting technique selection process but also for tuning the forecasting technique parameters (e.g., the weighting parameter allocation and the length of the time window in the $\text{WMA}_{n,w}$ – technique).

REFERENCES

Ahmadi, T. 2019. Inventory control systems with commitment lead time. PhD Thesis, Eindhoven University of Technology, Eindhoven.

Ahmadi, T., Z. Atan, T. de Kok, and I. Adan. 2019a. Optimal control policies for an inventory system with commitment lead time. *Naval Research Logistics* 66, no. 3: 193–212.

Ahmadi, T., Z. Atan, T. de Kok, and I. Adan. 2019b. Optimal control policies for assemble-to-order systems with commitment lead time. *IISE Transactions* 51, no. 12: 1365–1382.

Ahmadi, T., Z. Atan, T. de Kok, and I. Adan. 2020. Time-based service constraints for inventory systems with commitment lead time. *OR Spectrum* 42, no. 2: 355–395.

Ahmadi, T., M. Mahootchi, and K. Ponnambalam. 2018. Optimal randomized ordering policies for a capacitated two-echelon distribution inventory system. *Computers & Industrial Engineering* 124: 88–99.

Atan, Z., T. Ahmadi, C. Stegehuis, T. de Kok, and I. Adan. 2017. Assemble-to-order systems: A review. *European Journal of Operational Research* 261, no. 3: 866–879.

Bijvank, M., and Vis, I. F. 2011. Lost-sales inventory theory: A review. *European Journal of Operational Research* 215, no. 1: 1–13.

Brown, R. G. 1956. *Exponential Smoothing for Predicting Demand*. Cambridge, MA: Arthur D. Little Inc.

Brynjolfsson, E., T. Geva, and S. Reichman. 2016. Crowd-squared: Amplifying the predictive power of search trend data. *Management Information Systems Quarterly* 40, no. 4: 941–961.

Cachon, G., and C. Terwiesch. 2019. *Matching supply with demand: An introduction to Operations Management*. 4th Ed. New York: McGraw-Hill Education.

Chase, C. W. 2013. *Demand-Driven Forecasting: A Structured Approach to Forecasting.* 2nd Ed. Hoboken, NJ: John Wiley & Sons, Inc.

Chatfield, C. 1978. The Holt-Winters forecasting procedure. *Applied Statistics* 27, no. 3: 264–279.

Chatfield, D. C., and J. C. Hayya. 2007. All-zero forecasts for lumpy demand: A factorial study. *International Journal of Production Research* 45, no. 4: 935–950.

Chen, F., Z. Drezner, J. K. Ryan, and D. Simchi-Levi. 2000. Quantifying the bullwhip effect in a simple supply chain: The impact of forecasting, lead times, and information. *Management Science* 46, no. 3: 436–443.

Cleveland, W. P., and G. C. Tiao. 1976. Decomposition of seasonal time series: A model for the census X-11 program. *Journal of the American Statistical Association* 71, no. 355: 581–587.

Croston, J. D. 1974. Stock levels for slow-moving items. *Journal of the Operational Research Society* 25, no. 1: 123–130.

Croxton, K. L., D. M. Lambert, S. J. García-Dastugue, and D. S. Rogers. 2002. The demand management process. *International Journal of Logistics Management* 13, no. 2: 51–66.

Doganis, P., A. Alexandridis, P. Patrinos, and H. Sarimveis. 2006. Time series sales forecasting for short shelf-life food products based on artificial neural networks and evolutionary computing. *Journal of Food Engineering* 75, no. 2: 196–204.

Forrester, J. 1958. Industrial dynamics—a major breakthrough for decision-makers. *Harvard Business Review* 36, no. 4: 37–66.

Garcia, R. 2005. Uses of agent-based modeling in innovation/new product development research. *Journal of Product Innovation Management* 22, no. 5: 380–398.

Gardner, E. S., and E. Mckenzie. 1985. Forecasting trends in time series. *Management Science* 31, no. 10: 1237–1246.

Gligor, D. M. 2014. The role of demand management in achieving supply chain agility. *Supply Chain Management: An International Journal* 19, no. 5/6: 577–591.

Grekousis, G., and Y. Liu. 2019. Where will the next emergency event occur? Predicting ambulance demand in emergency medical services using artificial intelligence. *Computers, Environment and Urban Systems* 76: 110–122.

Grimson, J. A., and D. F. Pyke. 2007. Sales and operations planning: An exploratory study and framework. *International Journal of Logistics Management* 18, no. 3: 322–346.

Gutierrez, R. S., A. O. Solis, and S. Mukhopadhyay. 2008. Lumpy demand forecasting using neural networks. *International Journal of Production Economics* 111, no. 2: 409–420.

Güven, İ., and F. Şimşir. 2020. Demand forecasting with color parameter in retail apparel industry using artificial neural networks (ANN) and support vector machines (SVM) methods. *Computers & Industrial Engineering* 147: 106678.

He, Y., J. Jiao, Q. Chen, S. Ge, Y. Chang, and Y. Xu. 2017. Urban long term electricity demand forecast method based on system dynamics of the new economic normal: The case of Tianjin. *Energy* 133: 9–22.

Huber, J., and H. Stuckenschmidt. 2020. Daily retail demand forecasting using machine learning with emphasis on calendric special days. *International Journal of Forecasting* 36, no. 4: 1420–1438.

Hyndman, R. J., and A. B. Koehler. 2006. Another look at measures of forecast accuracy. *International Journal of Forecasting* 22, no. 4: 679–688.

Kremer, M., B. Moritz, and E. Siemsen. 2011. Demand forecasting behavior: System neglect and change detection. *Management Science* 57, no. 10: 1827–1843.

Lau, R. Y. K., W. Zhang, and W. Xu. 2018. Parallel aspect-oriented sentiment analysis for sales forecasting with big data. *Production and Operations Management* 27, no. 10: 1775–1794.

Law, R., G. Li, D. K. C. Fong, and X. Han. 2019. Tourism demand forecasting: A deep learning approach. *Annals of Tourism Research* 75: 410–423.

Liang, W.-Y., and C.-C. Huang. 2006. Agent-based demand forecast in multi-echelon supply chain. *Decision Support Systems* 42, no. 1: 390–407.

Lilien, G. L., P. Kotler, and K. S. Moorthy. 1992. *Marketing Models*. Englewood Cliffs, NJ: Prentice Hall, International Editions.

Mentzer, J. T., and M. A. Moon. 2004. Understanding demand. *Supply Chain Management Review* 8, no. 4: 38–45.

Mentzer, J. T., and M. A. Moon. 2005. Sales Forecasting management: A demand management approach. Sales forecasting management: A demand management approach. Thousand Oaks, CA: SAGE Publications.

Mielczarek, B., and J. Zabawa. 2017. Simulation model for studying impact of demographic, temporal, and geographic factors on hospital demand. 2017 Winter Simulation Conference (WSC), IEEE, Las Vegas, NV. 4498–4500.

Mirasgedis, S., Y. Sarafidis, E. Georgopoulou, D. Lalas, M. Moschovits, F. Karagiannis, and D. Papakonstantinou. 2006. Models for mid-term electricity demand forecasting incorporating weather influences. *Energy* 31, no. 2–3: 208–227.

Mohamed, Z., and P. Bodger. 2005. Forecasting electricity consumption in New Zealand using economic and demographic variables. *Energy* 30, no. 10: 1833–1843.

Mukhopadhyay, S., A. O. Solis, and R. S. Gutierrez. 2012. The accuracy of non-traditional versus traditional methods of forecasting lumpy demand. *Journal of Forecasting* 31: 721–735.

Petropoulos, F., S. Makridakis, V. Assimakopoulos, and K. Nikolopoulos. 2014. 'Horses for courses' in demand forecasting. *European Journal of Operational Research* 237, no. 1: 152–163.

Rawlings, J. O., S. G. Pantula, and D. A. Dickey. 2001. *Applied Regression Analysis: A Research Tool*. 2nd Ed. New York: Springer-Verlag.

Rexhausen, D., R. Pibernik, and G. Kaiser. 2012. Customer-facing supply chain practices-The impact of demand and distribution management on supply chain success. *Journal of Operations Management* 30, no. 4: 269–281.

Rosienkiewicz, M. 2020. Accuracy assessment of artificial intelligence-based hybrid models for spare parts demand forecasting in mining industry. In *Advances in Intelligent Systems and Computing*, 176–187. Cham, Switzerland: Springer.

Ryu, S., J. Noh, and H. Kim. 2016. Deep neural network based demand side short term load forecasting. *Energies* 10, no. 1: 3.

Schaer, O., N. Kourentzes, and R. Fildes. 2019. Demand forecasting with user-generated online information. *International Journal of Forecasting* 35, no. 1: 197–212.

Solaimani, S., E. Gulyaz, J. A. A. Van Der Veen, and V. Venugopal. 2015. Enablers and inhibitors of collaborative supply chains: An integrative framework. The 26th Conference of Production and Operations Management Society (POMS), Washington, DC.

Sun, S., Y. Wei, K. L. Tsui, and S. Wang. 2019. Forecasting tourist arrivals with machine learning and internet search index. *Tourism Management* 70: 1–10.

Suryani, E., S. Y. Chou, and C. H. Chen. 2012. Dynamic simulation model of air cargo demand forecast and terminal capacity planning. *Simulation Modelling Practice and Theory* 28: 27–41.

Syntetos, A., and J. Boylan. 2001. On the bias of intermittent demand estimates. *International Journal of Production Economics* 71, no. 1–3: 457–466.

Tako, A. A., and S. Robinson. 2012. The application of discrete event simulation and system dynamics in the logistics and supply chain context. *Decision Support Systems* 52, no. 4: 802–815.

Thomopoulos, N. T. 2015. Demand forecasting for inventory control. In *Demand Forecasting for Inventory Control*, 1–10. Cham: Springer International Publishing.

Thompson, P. A. 1990. An MSE statistic for comparing forecast accuracy across series. *International Journal of Forecasting* 6, no. 2: 219–227.

Trigg, D. W., and A. G. Leach. 1967. Exponential smoothing with an adaptive response rate. *Operations Research* 18, no. 1: 53.

Tsekouras, G. J., E. N. Dialynas, N. D. Hatziargyriou, and S. Kavatza. 2007. A non-linear multivariable regression model for midterm energy forecasting of power systems. *Electric Power Systems Research* 77, no. 12: 1560–1568.

Vulcano, G., G. van Ryzin, and R. Ratliff. 2012. Estimating primary demand for substitutable products from sales transaction data. *Operations Research* 60, no. 2: 313–334.

Winters, P. R. 1960. Forecasting sales by exponentially weighted moving averages. *Management Science* 6, no. 3: 324–342.

Wuest, T., D. Weimer, C. Irgens, and K.-D. Thoben. 2016. Machine learning in manufacturing: Advantages, challenges, and applications. *Production & Manufacturing Research* 4, no. 1: 23–45.

Yan, Q., C. Qin, M. Nie, and L. Yang. 2018. Forecasting the electricity demand and market shares in retail electricity market based on system dynamics and Markov chain. *Mathematical Problems in Engineering* 2018: 1–11.

Zhang, F. and R. Zhang. 2018. Trade-in remanufacturing, customer purchasing behavior, and government policy. *Manufacturing & Service Operations Management* 20, no. 4: 601–616.

Zhang, H., X. Zhang, and B. Zhang. 2009. System dynamics approach to urban water demand forecasting. *Transactions of Tianjin University* 15, no. 1: 70–74.

Zhang, L., and D. Levinson. 2004. Agent-based approach to travel demand modeling: Exploratory analysis. *Transportation Research Record: Journal of the Transportation Research Board* 1898, no. 1: 28–36.

Zhu, Z., B. Hoon Hen, and K. Liang Teow. 2012. Estimating ICU bed capacity using discrete event simulation. *International Journal of Health Care Quality Assurance* 25, no. 2: 134–144.

15 Conclusion

Farnaz Khoshgehbari and Mohsen S. Sajadieh
Department of Industrial Engineering, Amirkabir
University of Technology, Tehran, Iran

CONTENTS

15.1 INTRODUCTION

Supply chain management (SCM) refers to the global management of multiple relationships between different parts of the supply chain, from suppliers to final consumers. An effective supply chain employs eight key business processes to coordinate components of the supply chain in the best possible way to create added value in the system at the lowest cost. The eight key business processes are as follows: (1) customer relationship management, (2) customer service management, (3) demand management, (4) order fulfillment, (5) manufacturing flow management, (6) supplier relationship management, (7) product development, and (8) returns management (Lambert, Cooper, and Pagh 1998).

The most important goal of SCM is to reach maximum consumer satisfaction along with increasing profitability and productivity. For this purpose, demand management provides a structure to forecast customers' needs and synchronize supply chain capabilities (Croxton, Lambert et al. 2002). The demand management process includes strategic and operational steps. In the strategic step, demand management goals and strategies are determined, with considering information flow, forecasting and synchronizing procedures are developed and, finally, contingency management systems and the frameworks of metrics are developed. In the operational step, the data is collected, demand estimation and supply chain synchronization are performed and, finally, the performance of SCM is measured (Croxton et al. 2001).

Effective demand management increases customer loyalty, sales amounts and supply chain flexibility. Simultaneously, it decreases total expenses, safety stock, inventory level and demand variability (Gunasekaran, Patel, and Tirtiroglu 2001). Mentioned improvements are among the most important improvements considered for an efficient supply chain (Towill and McCullen 1999). Therefore, investing in demand management ultimately leads to greater supply chain efficiency and profitability.

As mentioned, demand management is about forecasting demand and synchronizing all parts of the supply chain to satisfy Customer demand. To increase demand or control its uncertainty and variability, influencing on demand will arise. There are many factors affecting demand that recognizing these factors is one of the most important measures of demand management. Identifying these factors, recognizing their impact, planning on and controlling them to improve supply chain efficiency can significantly help managers to create a productive supply chain.

The focus of this book is to introduce the factors influencing demand so that in each chapter of this book, one of the major factors influencing demand is thoroughly examined. Also, there are other factors that affect demand, but due to the lack of research background or limitations in writing this book, a separate chapter has not been assigned to them. These factors are briefly introduced in this chapter. The factors that have been examined in the book chapters are briefly introduced in Section 15.2 and additional factors are introduced in Section 15.3.

15.2 AN OVERVIEW OF THE BOOK CHAPTERS

As mentioned, the goal of this book, in general, is to comprehensively introduce the factors influencing demand. Several important factors are each introduced and surveyed in separate chapters. Chapters 2–13 of the book address these factors. Chapter 14 examines some functions and explains how to achieve them.

One of the most important goals of this conclusion is to provide a summary of the book chapters, which helps readers identify all the factors together. Readers are referred to the relevant chapter to study the details. The summary of the book chapters is as follows:

In Chapter 2, pricing as one of the most important factors in supply chains' profit, which can influence both demand and supply is investigated in detail. Deterministic and stochastic price-dependent demand functions are introduced, and the applications of pricing strategies in real situations are studied. Finally, future studies related to this issue are presented in different categories.

In Chapter 3, on-shelf inventory is introduced as an important factor influencing demand at the end of the supply chain where the final consumer is located because visible inventory is directly related to product availability from the customer's point of view, which affects product demand. Various inventory-dependent demand functions are introduced and optimal decisions are evaluated. Also, the relationships between some main factors and on-shelf inventory are examined and presented in detail. Finally, future studies in this area are presented.

In Chapter 4, rebate contract as a supply chain's essential strategic decision that directly affects customers' decisions and demand is investigated. The basic condition studied in the literature is two-echelon supply chain with a newsvendor model. The top five most common types of rebates used in the literature are mail-in rebate, manufacturer rebate, retailer rebate, consumer rebate and wholesale price rebate. Deterministic, hybrid, conditional, and Poisson functions are studied. Also, considering the uncertainties of the real world, uncertain models have been presented in this field. In a supply chain, there is always competition between the manufacturer and the retailer for more profit. Game theory provides a win-win situation for both firms to make suitable profits, so the game theory is mostly used as a solution which is thoroughly investigated.

In Chapter 5, the effects of service level and its applications in different demand functions are analyzed. Focusing on the related literature, the most common linear and non-linear service level–dependent demand functions in four distinct categories are also described. The dependent demand functions are presented as service level–dependent, quality-dependent, lead time-dependent and reliability-dependent forms. In addition, the applications of the introduced demand functions are investigated in real situations to obtain a more comprehensive view. Following the research trend and optimization models, the main insights are presented for future practitioners to know how they can apply service level in supply chain or revenue management problems and evaluate service level effects on customer demand more appropriately.

In Chapter 6, advertising and marketing as important tools in increasing demand are investigated. Advertising and marketing decisions in the field of operations management have been considered by many researchers. In reviewed articles, the effect of advertising and marketing efforts on demand functions in forms of power, square root, linear and non-linear with game theory approaches and in the form of static/dynamic models with deterministic/stochastic answers have been studied. Also, the issues of contract design (cooperative advertising, participation in costs related to marketing efforts, etc.) to coordinate the supply chain are among the topics of interest in this area. Also, advertising and marketing efforts in B2B, B2C and closed-loop supply chains are studied.

In Chapter 7, the effect of the company's reputation on the company's competitive advantage and performance is examined. Company's reputation affects stakeholders and customers' behavior and needs, and ultimately it affects demand. Some mathematical models are introduced to investigate reputation-dependent demand functions. With introducing and reviewing several case studies, the advantages and disadvantages of these functions are investigated and, finally, by identifying study gaps, suggestions for future research in this field are provided.

In Chapter 8, different approaches to individual customer choice behavior modeling in congested systems are reviewed. There are also some factors influencing customer choice behavior in queueing systems. These factors are introduced and studied in detail.

In Chapter 9, dynamic innovation capabilities are investigated as important tools to increase company's competitive advantage. Leading in innovation and applying it have a direct impact on customer's behavior and company's reputation, which in turn affect demand. Several case studies are presented and, the decision-making effects in this field on the supply chain are investigated and identified.

In Chapter 10, time-dependent demand functions are surveyed for better understanding of different functions' applicability and features. Accordingly, the functions are grouped into four principal ones, namely, linear time-dependent demand, exponential time-dependent demand, quadratic time-dependent demand and, ramp-type time-dependent demand. After the survey, a comprehensive literature review is provided to shed light on future research directions. Also, two real case studies are examined to show the functions' applications and, the mathematical models with their solution approaches are presented.

In Chapter 11, inflation as a factor that influences demand is investigated. Some practical examples and case studies are mentioned. Different forms of demand functions are introduced and suggestions for future studies in this field are provided. It is mentioned that inflation could be divided into two categories, demand pull inflation and cost push inflation. Since the nature and effects of inflation are uncertain and demand is one of the most important factors affecting the supply chain, investigating the dynamic interaction between inflation and demand is important for supply chain management and especially the inventory system. Models that have examined the impact of inflation on the inventory system are introduced and reviewed.

In Chapter 12, the effects of pricing and stock level of substitute and complementary products are investigated. Studying the interaction between demand of complementary and substitute products helps to optimize the decision-making process, identify the opportunities and increase marginal profit. The linear demand function has been widely used in the literature because processes and calculations on the linear demand curve are easier and more understandable. This linear function is introduced and investigated in detail.

In Chapter 13, it is investigated how search costs may discourage consumers from buying and may result in the incompetent allocating of products and resources since customers may get what they do not like. Although search cost for information of products and price has been reduced, the high search cost of quality may lead to customer back from purchasing. This chapter studies previous researches and discusses the factors of social networks that affect demand. Moreover, both quantitative and qualitative attributes of online reviews are considered. It is shown that when customers read online reviews, they pay attention to other information, such as reviewers' reputation, in addition to the review scores.

Finally, in Chapter 14, forecasting the independent-demand models and estimating the parameters of the dependent-demand models are investigated. Time-series forecasting techniques have been classified for forecasting the independent-demand models and, regression and simulation-based models have been employed for estimating the parameters of the dependent-demand models. Some methods are provided to measure the accuracy of the historical data and, finally, Big Data and artificial intelligence (AI) are mentioned as methods for improving the demand forecasting process.

15.3 OTHER FACTORS INFLUENCING DEMAND

It should be noted that there are many factors affecting demand. They have different effects in different countries, conditions, cultures and even climates. In addition to the factors investigated in the previous chapters, there are other factors that have a large impact on demand. Although many factors are still unknown, in Chapter 15 we have tried to present as many additional factors as possible.

15.3.1 PRODUCT QUALITY

Product quality is one of the most important factors in customer satisfaction and loyalty. This factor plays an important role in all parts of the supply chain and affects both B2B and B2C relations. Vörös (2002) has introduced the concept of quality inflation, which means customer expectations increase over time due to their experience of high-quality products. These expectations make companies increase their products' quality and operational efficiency to hold their market share and competitive advantages. However, improving the quality increases the total expenses and the price of goods and, ultimately, it reduces the marginal profit of the manufacturer, but demand can be increased by improving the quality level at any price.

Maiti and Giri (2014) have assumed a positive linear relationship between quality and demand in a closed-loop supply chain and have concluded that product quality has a positive impact on demand rate and increases the supply chain's profit. Giovanni (2011) has considered customer demand as a linear function of wholesale price and goodwill where goodwill has a positive linear relationship with quality and advertising. Vörös (2002) has presented a non-linear relationship between demand and quality. Hallak (2006) has investigated the relationship between the product's quality and the direction of trade and has concluded that rich countries tend to import more from countries that produce high-quality products.

15.3.2 SOCIAL RESPONSIBILITY

There are many reasons why a company should be socially responsible, such as ecological issues (Bansal and Roth 2000) and financial performance (Wang, Dou and Jia 2015). Given the extensive literature, corporate social responsibility (CSR) is one of the best operations a company can use to increase performance and profitability (M.-D. P. Lee 2008, McWilliams and Siegel 2001). For example, due to climate changes, one of the most important actions a firm can take to get a positive response from the market is reducing carbon emissions (Lee, Park, and Klassen 2015). Albuquerque, Koskinen, and Zhang (2019) have studied the relationship between CSR and firms' systematic risk based on the premise that CSR is a product differentiation strategy. Raza (2018) has generated a quantitative model for joint pricing, retailer inventory ordering and investment for socially responsible decisions in a supply chain under both deterministic and stochastic demand.

15.3.3 DISTANCE (FACILITY LOCATION)

The distance of one point from the production sites may increase the cost of transportation and consequently the price of the goods, which, depending on the type of goods, can affect the demand. Also, goods supplied from abroad have a higher price elasticity at the country's borders than in the center of the country (Asplund, Friberg, and Wilander 2007).

15.3.4 THE SIZE AND THE STRUCTURE OF THE POPULATION

Like many factors that affect social trends, population structure has a great impact on needs, trends and demand structure. Population aging over time increases inflation and unemployment and, it decreases gross domestic product (GDP) (Katagiri 2012). Age distribution of the population has important effects on house demand (Ermisch 1996). Also, it is obvious that a country with a high youth-to-middle-age ratio has a greater demand for high-tech products.

15.3.5 CLIMATE CONDITIONS

With the increase of industrial activities and the effect of greenhouse gases on the environment, the earth is warming day by day, and this affects the climatic conditions of different regions over time. Agriculture, food, tobacco and lumber industries with electricity and gas services are significantly influenced by climate change. Although the effects are not uniform from one country to another (Mendelsohn et al. 2000), the low impact of climate change on mortality and morbidity rates has a great impact on people's welfare and population structure, which in turn can change social behaviors and customer needs (Jorgenson et al. 2004).

A specific example for the impact of climate change on demand is heating and cooling devices. As the earth gets warmer, electricity demand for cooling increases and electricity demand for heating decreases, so the demand of their related products will be changed. Besides, the demand for heat sources, including gas and electricity decreases due to rising temperatures (Mideksa and Kallbekken 2010).

15.3.6 GOVERNMENT POLICIES

The government plays a major role in creating culture, changing social behaviors by enforcing do's and don'ts, enacting tax laws and advocacy strategies. For example, government policies that support the production of green foods will increase their demand and improve people's health and well-being over time. Also, in some countries, the consumption of some products is prohibited, such as Muslim countries that ban the consumption of alcohol or banning the smoking in some countries, these policies deter people from buying and consuming illicit goods and therefore affect demand.

15.3.7 INCOME

Hicks (1946), has investigated the effects of income on customer behaviors from an economic perspective. First, he refers to Marshall's theory, which assumes the demand rate of a commodity depends on the marginal utility of money, that is, a fixed parameter for an individual; therefore, in Marshall's theory, without changes in prices, even if an individual income increases, demand will not change. Contrary to this simplification, changes in the composition of the consumption of goods and demand due to changes in income, for both the individual and the outcome of the individuals who make up society, is obvious. Changes in income have different effects on the demand of different types of products, and it is necessary to mention that it differs for different segments of the population (Hicks 1946).

15.3.8 INTERNATIONAL MARKETS

One of the consequences of globalization is the genesis of international markets. International markets can change consumer's needs and behaviors by introducing new products, bringing technology to the country and increasing the competitiveness of domestic producers (Hellwig 2014). Also, attending at international markets increases the reputation and profitability of companies, which in turn has a significant impact on demand (Brander and Spencer 1985).

15.3.9 CRITICAL CONDITIONS, SUCH AS FAMINE, WAR, AND THE EPIDEMIC OF DISEASE

Critical conditions in a country change the behaviors and concerns of the people. For example, in a country that is at war, many people are no longer willing to buy luxury goods, and in this situation, the demand for basic goods and even Giffen goods increases. Also, in an epidemic of a disease, the demand for related health goods and health services increases. It must be noted that impacts vary among different countries, conditions and products.

15.3.10 UNEMPLOYMENT RATE

Unemployment causes lack of income and inability to cover expenses. As mentioned earlier, income has an important effect on social welfare and, consequently, on social behaviors or even culture (Ravallion and Lokshin 2001). In a society with a high unemployment rate, a large percentage of people are below the poverty line and therefore need government assistance, in this situation, their needs are limited to the basic needs of food and clothing, which affects the demand of other products.

15.3.11 SOCIAL BEHAVIORS

Social behaviors have a very important effect on demand for water and energy (Faiers, Matt Cook, and Neame 2007, Koutiva and Makropoulos 2016). Also, there are some special social behaviors, such as concerns about halal products that affect Muslim consumers' willingness to pay (WTP) and the demand for halal foods in Muslim's countries (Ahmed et al. 2019). Besides, due to increasing environmental issues, young generations seem to be more inclined to use green and environmentally friendly products (Kanchanapibul et al. 2014). From the companies' point of view, social marketing, motivational interviewing and community mobilization are related to product awareness, community adoption with a new product, social trends and, finally, social behaviors (Quick 2003, Lewis 2005).

15.3.12 THE LENGTH OF THE CREDIT PERIOD

Late payment allows the retailer to think about its own benefits and the customer's satisfaction without worrying about paying for the custom product. In other words, the retailer uses its credit and delays the payment. The credit period, by increasing flexibility, increases the retailers' desire to place more orders, creates a long-term interaction between parts of the supply chain and, ultimately, increases the quality of products by creating a time interval between delivery and receipt (Yang, Hong, and Lee 2014). It should be noted that this factor is briefly mentioned in Chapters 3 and 4 of this book.

The credit period has been examined from two aspects of finance and operations management (Jing, Chen, and Cai 2012). From an operations management point of view, the relevant literature can be divided into two parts, the supply side and the demand side. The most recent researches on this subject have considered the credit period as a coordination tool that deals with how to decide on credit, the duration of credit and the settlement period.

As mentioned earlier, the factors presented in this book are not all factors that affect demand. Many factors are still unknown and, many others have indirect effects on demand. The authors of this book have tried to identify and present the factors as much as possible. The investigated factors can be divided into two categories: endogenous factors and exogenous factors. Figure 15.1 shows all these factors at a glance.

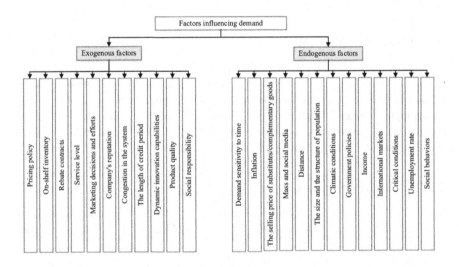

FIGURE 15.1   Factors influencing demand at a glance.

As mentioned, influencing demand is one of the most important demand management strategies in supply chain management. Identifying the factors influencing demand is the first step to manage it. In this book, all the major factors affecting demand have been introduced and investigated as much as possible with mathematical and optimization point of view in the chapters. Also, several additional factors have been introduced and investigated. Further study on the impact of these factors on demand from an optimization perspective can be a good suggestion for future studies in the field of influencing demand.

REFERENCES

Albuquerque, Rui, Yrjö Koskinen, and Chendi Zhang. 2019. Corporate social responsibility and firm risk: Theory and empirical evidence. *Management Science* 65, no. 10: 4451–4469.

Ahmed, Waqar, Arsalan Najmi, Hafiz Muhammad Faizan, and Shaharyar Ahmed. 2019. Consumer behaviour towards willingness to pay for Halal products: An assessment of demand for Halal certification in a Muslim country. *British Food Journal* 121, no. 2: 492–504.

Asplund, Marcus, Richard Friberg, and Fredrik Wilander. 2007. Demand and distance: Evidence on cross-border shopping. *Journal of Public Economics* 91, no. 1–2: 141–157.

Bansal, Pratima, and Kendall Roth. 2000. Why companies go green: A model of ecological responsiveness. *Academy of Management Journal* 43, no. 4: 717–736.

Brander, James A., and Barbara J. Spencer. 1985. Export subsidies and international market share rivalry. *Journal of International Economics* 18, no. 1–2: 83–100.

Croxton, Keely L., Sebastián J. García-Dastugue, Douglas M. Lambert, and Dale S. Rogers. 2001. The supply chain management processes. *International Journal of Logistics Management* 12, no. 2: 13–36.

Croxton, Keely L., Douglas M. Lambert, Sebastián J. García-Dastugue, and Dale S. Rogers. 2002. The Demand Management Process. *International Journal of Logistics Management* 13, no. 2: 51–66.

Ermisch, John. 1996. The demand for housing in britain and population ageing: Microeconometric evidence. *Economica* 63, no. 251: 383–404.

Faiers, Adam, Matt Cook, and Charles Neame. 2007. Towards a contemporary approach for understanding consumer behaviour in the context of domestic energy use. *Energy Policy* 35, no. 8: 4381–4390.

Giovanni, Pietro De. 2011. Quality improvement vs. advertising support: Which strategy works better for a manufacturer? *European Journal of Operational Research* 208, no. 2: 119–130.

Gunasekaran, A., C. Patel, and E. Tirtiroglu. 2001. Performance measures and metrics in a supply chain environment. *International Journal of Operations & Production Management* 21, no. 1/2: 71–87.

Hallak, Juan Carlos. 2006. Product quality and the direction of trade. *Journal of International Economics* 68: 238–265.

Hellwig, Timothy. 2014. Balancing Demands: The world economy and the composition of policy preferences. *Journal of Politics* 76, no. 1: 1–14.

Hicks, John R. 1946. *Value and Capital: An Inquiry into Some Fundamental Principles of Economic Theory*. 2nd Ed. London: Oxford: Clarendon Press.

Jing, Bing, Xiangfeng Chen, and Gangshu (George) Cai. 2012. Equilibrium financing in a distribution channel with capital constraint. *Production and Operations Management* 21, no. 6: 1090–1101.

Jorgenson, Dale W., Richard J. Goettle, Brian H. Hurd, and Joel B. Smith. 2004. *U.S. Market Consequences of Global Climate Change*. Arlington, United States: Pew Center on Global Climate Change.

Kanchanapibul, Maturos, Ewelina Lacka, Xiaojun Wang, and Hing Kai Chan. 2014. An empirical investigation of green purchase behaviour among the young generation. *Journal of Cleaner Production* 66: 528–536.

Katagiri, Mitsuru. 2012. Economic Consequences of Population Aging in Japan: Effects through Changes in Demand Structure. Institute for Monetary and Economic Studies (IMES) Discussion Paper Series 12-E-03. Institute for Monetary and Economic Studies, Bank of Japan.

Koutiva, Ifigeneia, and Christos Makropoulos. 2016. Modelling domestic water demand: An agent based approach. *Environmental Modelling & Software* 79: 35–54.

Lambert, Douglas M., Martha C. Cooper, and Janus D. Pagh. 1998. Supply chain management: implementation issues and research opportunities. *International Journal of Logistics Management* 9, no. 2: 1–20.

Lee, Min-Dong Paul. 2008. A review of the theories of corporate social responsibility: Its evolutionary path and the road ahead. *International Journal of Management Reviews* 10, no. 1: 53–73.

Lee, Su-Yol, Yun-Seon Park, and Robert D. Klassen. 2015. Market responses to firms' voluntary climate change information disclosure and carbon communication. *Corporate Social Responsibility and Environmental Management* 22, no. 1: 1–12.

Lewis, Michael. 2005. Incorporating strategic consumer behavior into customer valuation. *Journal of Marketing* 69, no. 4: 230–238.

Maiti, T., and B. C. Giri. 2014. A closed loop supply chain under retail price and product quality dependent demand. *Journal of Manufacturing Systems* 37: 624–637.

McWilliams, Abagail, and Donald Siegel. 2001. Corporate social responsibility: A theory of the firm perspective. *Academy of Management Review* 26, no. 1: 117–127.

Mendelsohn, Robert, Wendy Morrison, Michael E. Schlesinger, and Natalia G. Andronova. 2000. Country-specific market impacts of climate change. *Climatic Change* 45, no. 3–4: 553–569.

Mideksa, Torben K., and Steffen Kallbekken. 2010. The impact of climate change on the electricity market: A review. *Energy Policy* 38, no. 7: 3579–3585.

Quick, R. 2003. Changing community behaviour: Experience from three African countries. *International Journal of Environmental Health Research* 13: S115–S121.

Ravallion, Martin, and Michael Lokshin. 2001. Identifying welfare effects from subjective questions. *Economica* 68, no. 271: 335–357.

Raza, Syed Asif. 2018. Supply chain coordination under a revenue-sharing contract with corporate social responsibility and partial demand information. *International Journal of Production Economics* 205: 1–14.

Towill, Denis R., and Peter McCullen. 1999. The impact of agile manufacturing on supply chain dynamics. *International Journal of Logistics Management* 10, no. 1: 83–96.

Vörös, József. 2002. Product balancing under conditions of quality inflation, cost pressures and growth strategies. *European Journal of Operational Research* 141, no. 1: 153–166.

Wang, Qian, Junsheng Dou, and Shenghua Jia. 2015. A meta-analytic review of corporate social responsibility and corporate financial performance: The moderating effect of contextual factors. *Business & Society* 55, no. 8: 1083–1121.

Yang, Shuai, Ki-sung Hong, and Chulung Lee. 2014. Supply chain coordination with stock-dependent demand rate and credit incentives. *International Journal of Production Economics* 157, no. 1: 105–111.

Index

Note: Page numbers in *italics* indicate a figure.

Printed in the United States
by Baker & Taylor Publisher Services